HOW TO PASS ✓

PRACTICE & PREPARATION

HIGHER

MATHS

Brian J Logan

**HODDER
GIBSON**
AN HACHETTE UK COMPANY

Although every effort has been made to ensure that website addresses are correct at time of going to press, Hodder Gibson cannot be held responsible for the content of any website mentioned in this book. It is sometimes possible to find a relocated web page by typing in the address of the home page for a website in the URL window of your browser.

Hachette UK's policy is to use papers that are natural, renewable and recyclable products and made from wood grown in sustainable forests. The logging and manufacturing processes are expected to conform to the environmental regulations of the country of origin.

Orders: please contact Bookpoint Ltd, 130 Milton Park, Abingdon, Oxon OX14 4SB. Telephone: (44) 01235 827720. Fax: (44) 01235 400454. Lines are open 9.00–5.00, Monday to Saturday, with a 24-hour message answering service. Visit our website at www.hoddereducation.co.uk. Hodder Gibson can be contacted direct on: Tel: 0141 848 1609; Fax: 0141 889 6315; email: hoddergibson@hodder.co.uk

© Brian Logan 2011
First published in 2011 by
Hodder Gibson, an imprint of Hodder Education,
An Hachette UK Company
2a Christie Street
Paisley PA1 1NB

Impression number   5 4 3 2 1
Year   2014 2013 2012 2011

Cover photo © Pawel Gaul / iStockphoto; running head image © Photodisc/Getty Images
Illustrations by DC Graphic Design Limited, Swanley Village, Kent.
Typeset in Plantin 13pt by DC Graphic Design Limited, Swanley Village, Kent.
Printed in the UK by CPI Antony Rowe

A catalogue record for this title is available from the British Library

ISBN: 978 1444 108385

Contents

Foreword

In addition to the general disclaimer regarding SQA endorsement on the inner front cover of this title, and in recognition of the very specific requirements for SQA Higher Maths papers, the publishers wish to clarify that this book is not intended to be a substitute for official SQA past papers, whose equality from year to year requires to be guaranteed, and where the proportion and level of questions is strictly monitored within each individual paper. Although the coverage of units and competency levels within individual papers of this book is not what it would be in an actual Higher Maths exam, the book contains an overall quota of Unit coverage that would be in line with a single paper requirement (at least 25% from each of the three Units) but deliberately does not attempt to do so rigidly within each paper. Paper A, for example, leans more towards Unit I, which allows students valuable exam-style practice from an early stage, but it should be noted that this structure would not be appropriate in the composition of a preliminary examination to be used as the basis for an appeal.

Although the objective test questions in the book include a range of levels covering C to A, and offer valuable advice on working methods to approach them, they have not been taken from SQA item banks, nor do they necessarily follow the typical pattern found in SQA Higher Maths Paper 1 Section A, where the first 16 or 17 questions would be at level C, later questions at levels A/B. Also, they have not necessarily been constructed in the same manner as the majority of SQA objective test questions where students have to make two mistakes to reach one of the incorrect answers.

Introduction

The purpose of this book is to help you prepare properly for your Higher Maths examination by giving you the opportunity to practise four specimen papers.

These papers are designed so that they follow closely the Higher Maths syllabus and reflect the style and content of the examination. Detailed solutions are given to every question. If, as you study the questions and their solutions, you cannot understand any of the solutions, it is important that you ask another student or your teacher to clarify the problem.

In addition to the practice papers and solutions, there are several shorter appendices later in the book. These give advice on useful facts and formulae, how to tackle objective test questions and offer some useful hints. You may decide it is helpful to read these appendices *before* you practise the practice papers. That decision is yours to make.

If you are reading this book early in the session, you will not be ready to attempt entire practice papers. In that case you may wish to practise questions from the topic you have been studying recently. For example, suppose you have recently completed the topic **The Straight Line**. If you wish to practise straight line questions, look at Appendix 4 which gives a list of all questions in the specimen papers topic by topic. One thing is sure, the more you practise examination papers, the more your maths will improve and with it your chances of not only passing, but getting a top grade.

So good luck!

Other Recommended Reading

How to Pass Higher Maths Colour Edition by Peter Westwood gives a very detailed and comprehensive guide to the Higher course, topic by topic, including a guide to revision techniques and advice on sitting the exam itself.

How to Pass Flash Revise: Higher Maths by Brian J Logan is a pocketbook with short question and answers covering the Higher syllabus. It is ideal for instant revision.

Introduction to Paper One (Non-calculator)

In this section, there are four complete Paper One question papers.

In Paper One, you are **not** allowed to use a calculator.

In Paper One, there are two sections. Section A consists of 20 objective test questions worth 2 marks each. In these questions, four possible answers, labelled A, B, C and D, are given. Only **one** of these answers is correct. Section B contains some longer questions and is worth 30 marks. In total Paper One is worth 70 marks. The time allocated to complete Paper One is 1 hour 30 minutes.

There is a set of solutions for each practice paper. These will include some hints and advice on how to tackle particular questions.

In all papers, you are allowed to use a list of formulae. You will find this list on the following page.

Each SQA Paper One begins with the following instructions:

Read carefully

Calculators may NOT be used in this paper.

Section A – Questions 1–20 (40 marks)

Instructions for completion of **Section A** are given on page two.

For this section of the examination you must use an **HB pencil**.

Section B (30 marks)

1 Full credit will be given only where the solution contains appropriate working.

2 Answers obtained by readings from scale drawings will not receive any credit.

LIST OF FORMULAE

Circle:

The equation $x^2 + y^2 + 2gx + 2fy + c = 0$ represents a circle with centre $(-g, -f)$ and radius $\sqrt{g^2 + f^2 - c}$.

The equation $(x-a)^2 + (y-b)^2 = r^2$ represents a circle with centre (a, b) and radius r.

Scalar Product: $\mathbf{a}.\mathbf{b} = |\mathbf{a}||\mathbf{b}|\cos\theta$, where θ is the angle between $\mathbf{a}$ and $\mathbf{b}$

$$\text{or} \quad \mathbf{a}.\mathbf{b} = a_1b_1 + a_2b_2 + a_3b_3 \text{ where } \mathbf{a} = \begin{pmatrix} a_1 \\ a_2 \\ a_3 \end{pmatrix} \text{ and } \mathbf{b} = \begin{pmatrix} b_1 \\ b_2 \\ b_3 \end{pmatrix}.$$

Trigonometric formulae:

$$\sin(A \pm B) = \sin A \cos B \pm \cos A \sin B$$

$$\cos(A \pm B) = \cos A \cos B \mp \sin A \sin B$$

$$\sin 2A = 2\sin A \cos A$$

$$\cos 2A = \cos^2 A - \sin^2 A$$

$$= 2\cos^2 A - 1$$

$$= 1 - 2\sin^2 A$$

Table of standard derivatives:

$f(x)$	$f'(x)$
$\sin ax$	$a\cos ax$
$\cos ax$	$-a\sin ax$

Table of standard integrals:

$f(x)$	$\int f(x)dx$
$\sin ax$	$-\dfrac{1}{a}\cos ax + C$
$\cos ax$	$\dfrac{1}{a}\sin ax + C$

4

Practice Paper A

PAPER ONE

Do not use a calculator

SECTION A

ALL questions should be attempted

1. A circle has equation $x^2 + y^2 + 10x - 4y - 35 = 0$.

   What is the radius of this circle?

   A $\sqrt{151}$

   B $\sqrt{56}$

   C   9

   D   8

2. If $f(x) = 2x^3 - 4$, find the value of $f'(2)$.

   A   20

   B   24

   C   12

   D   144

3. The straight line joining the points (6, 0) and (0, 12) passes through the point $(-2, a)$. Find the value of a.

   A   16

   B   7

   C   8

   D   20

4. A sequence is defined by the recurrence relation $u_{n+1} = 0 \cdot 1u_n + 28$ with $u_1 = 20$.

   What is the value of u_4?

   A   28·3

   B   28·4

   C   30·8

   D   31·1

5. If the value of $\sin x$ is $\dfrac{1}{3}$, where x is an acute angle, find the exact value of $\sin 2x$.

A $\quad -\dfrac{4\sqrt{2}}{9}$

B $\quad \dfrac{7}{9}$

C $\quad -\dfrac{7}{9}$

D $\quad \dfrac{4\sqrt{2}}{9}$

6. What is the derivative of $3x^2 + \dfrac{1}{x}$, where $x \neq 0$?

A $\quad 6x + \dfrac{1}{x^2}$

B $\quad 6x - \dfrac{1}{x^2}$

C $\quad 3x + 1$

D $\quad 6x$

7. What is the gradient of the straight line perpendicular to the line $x - 5y + 9 = 0$?

A $\quad -9$

B $\quad 5$

C $\quad \dfrac{1}{5}$

D $\quad -5$

8. A circle has the line joining A (2, 4) to B (12, 4) as a diameter.

What is the equation of the circle?

A $\quad (x-7)^2 + (y-4)^2 = 5$

B $\quad (x-10)^2 + y^2 = 25$

C $\quad (x-7)^2 + (y-4)^2 = 25$

D $\quad (x-10)^2 + y^2 = 5$

9. A function is given by $f(x) = 7x^2 + 4x + 1$.

Which of the following describes the nature of the roots of $f(x) = 0$?

A Two real roots

B Three real roots

C Two equal roots

D No real roots

10. Which of the following graphs could represent the function $f(x) = \log_3 x$?

A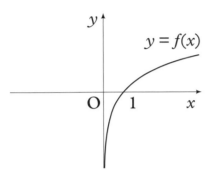

B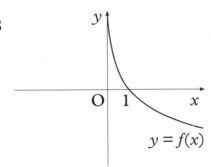

C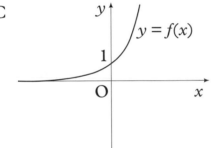

D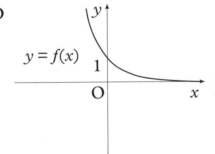

11. For what values of x is $x^2 - 2x - 15 > 0$?

A $x > 5$ only

B $x < -3$ only

C $-3 < x < 5$

D $x < -3, x > 5$

12. A sequence is generated by the recurrence relation $u_{n+1} = 0 \cdot 4u_n + 12$.

What is the limit of this sequence as $n \to \infty$?

A   20

B   30

C $\dfrac{60}{7}$

D $\dfrac{1}{20}$

13. A function is defined by $f(x) = 4\sin x° - 3\cos x°$.

What is the minimum value of this function?

A -12

B   5

C -5

D $\dfrac{3}{4}$

14. What is the exact value of $\sin\dfrac{7\pi}{6}$?

A $-\dfrac{\sqrt{3}}{2}$

B $-\dfrac{1}{2}$

C $\dfrac{1}{2}$

D $\dfrac{\sqrt{3}}{2}$

15. Here are two statements about the vector joining points A (3, –5, 7) and B (2, 0, 8).

(1) The vector $\overrightarrow{AB}$ has components $\begin{pmatrix} -1 \\ 5 \\ 1 \end{pmatrix}$.

(2) The length of $\overrightarrow{AB} = \sqrt{27}$.

Which of the following is true?

A   Neither statement is correct.

B   Only statement (1) is correct.

C   Only statement (2) is correct.

D   Both statements are correct.

16. A curve has equation $y = 2x^3 - 4x$.

What is the gradient of the tangent to the curve at the point (−2, −8)?

A   −8

B   8

C   20

D   −20

17. What is the smallest positive value of x which satisfies the equation
$\cos 4x° \cos x° + \sin 4x° \sin x° = \dfrac{1}{2}$?

A   12

B   20

C   30

D   60

18. The line EF makes an angle of $\dfrac{\pi}{3}$ radians with the *y*-axis, as shown in the diagram.

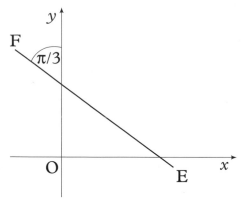

What is the gradient of EF?

A $\quad -\dfrac{1}{\sqrt{3}}$

B $\quad -\dfrac{1}{2}$

C $\quad -\dfrac{\sqrt{3}}{2}$

D $\quad -\sqrt{3}$

19. Given that $f(x)=\sqrt{(7-5x^2)}$ on a suitable domain, find $f'(x)$.

A $\quad \dfrac{-5x}{\sqrt{(7-5x^2)}}$

B $\quad \dfrac{1}{2\sqrt{(7-5x^2)}}$

C $\quad \dfrac{2}{\sqrt{(7-5x^2)}}$

D $\quad \dfrac{5x}{\sqrt{(7-5x^2)}}$

20. The graphs of $y = x^2 - 4$ and $y = 4 - x^2$ are shown in the diagram.

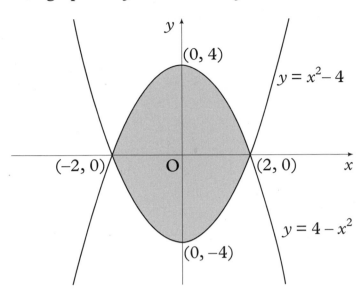

Which of the following gives the area of the shaded section?

A $\displaystyle\int_{-4}^{4} (8 - 2x^2)\,dx$

B $\displaystyle\int_{-2}^{2} (8 - 2x^2)\,dx$

C $\displaystyle\int_{-2}^{2} (2x^2 - 8)\,dx$

D $\displaystyle\int_{-4}^{4} (2x^2 - 8)\,dx$

[END OF SECTION A]

SECTION B

ALL questions should be attempted

21. The diagram shows a sketch of the function $y = f(x)$.

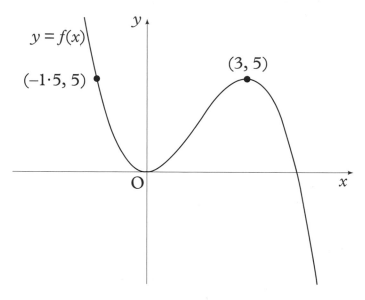

Copy the diagram and on it sketch the graph of $y = 2 - f(x)$. **(3)**

22. The vertices of a triangle are P (4, 7), Q (−3, 0) and R (5, −2). The point M is the midpoint of PR.

(a) Calculate the coordinates of M. **(1)**

(b) Find the equation of the median QM. **(2)**

(c) Find the equation of the altitude from R to PQ. **(3)**

(d) The median and altitude meet at T.
Find the coordinates of T. **(3)**

23. Find the coordinates of the turning points of the curve with equation

$y = x^3 - 3x^2 - 24x + 5$ and determine their nature. **(8)**

24. (a) By writing $3x$ as $(2x + x)$ and expanding, prove that

$$\cos 3x = 4\cos^3 x - 3\cos x.$$ **(4)**

(b) Make $\cos^3 x$ the subject of this formula. **(1)**

(c) The following diagram shows the graph of $y = \cos^3 x$ from $x = 0$ to $x = \dfrac{\pi}{2}$.

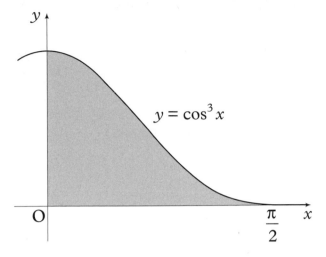

Hence or otherwise, calculate the area which has been shaded.   **(5)**

[END OF SECTION B]

[END OF QUESTION PAPER]

Practice Paper B

PAPER ONE

Do not use a calculator

SECTION A

ALL questions should be attempted

1. A sequence is defined by the recurrence relation

   $u_{n+1} = 0 \cdot 4 u_n + 8$ with $u_7 = 20$.

   What is the value of u_9?

   A $8 \cdot 8$

   B $10 \cdot 7$

   C $11 \cdot 2$

   D $14 \cdot 4$

2. What is the centre of the circle with equation $x^2 + y^2 + 8x - 10y - 15 = 0$?

   A $(-5, 4)$

   B $(4, -5)$

   C $(-4, 5)$

   D $(5, -4)$

3. The line $3y = 2x + 12$ meets the x-axis at P. What is the gradient of the line joining P to Q $(3, -5)$?

   A -3

   B $-\dfrac{5}{9}$

   C $\dfrac{5}{3}$

   D 5

4. Three collinear points have coordinates P (-5, -1, 2), Q (-1, 1, 8) and R (7, 5, 20) and Q lies between P and R.

    What is the ratio in which Q divides PR?

    A $1:1$

    B $1:2$

    C $1:4$

    D $1:6$

5. Find $\int (2x+1)^2 dx$.

    A $2(2x+1)+C$

    B $\dfrac{1}{4}(2x+1)^2 +C$

    C $\dfrac{1}{6}(2x+1)^3 +C$

    D $\dfrac{1}{3}(2x+1)^3(x^2+x)+C$

6. Functions f and g are defined on the set of real numbers by $f(x)=1-2x$ and $g(x)=x^2$.

    Find the value of $f(g(-1))$.

    A   9

    B   3

    C   1

    D -1

7. The x-axis is a tangent to a circle with centre $(-5,-8)$ as shown in the diagram.

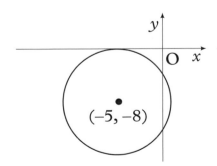

What is the equation of the circle?

A $(x+5)^2 + (y+8)^2 = 9$

B $(x+5)^2 + (y+8)^2 = 25$

C $(x-5)^2 + (y-8)^2 = 64$

D $(x+5)^2 + (y+8)^2 = 64$

8. A is the point $(4, d, 8)$ and B is $(-6, 10, e)$.

M $(-1, 4, -2)$ is the midpoint of AB.

What are the values of d and e?

	d	e
A	-2	-12
B	-2	12
C	-6	-12
D	-6	12

NOTE: There is usually one objective question in this style each year.

9. A sequence is generated by the recurrence relation $u_{n+1} = 0 \cdot 8u_n - 80$.

What is the limit of this sequence as $n \to \infty$?

A -400

B -100

C 100

D 400

10. Here are two statements about the equation $x^2 + 2px + p^2 = 0$,

where $p>0$, p is real:

(1) The roots are real. (2) The roots are equal.

Which of the following is true?

A Neither statement is correct.

B Only statement (1) is correct.

C Only statement (2) is correct.

D Both statements are correct.

11. The diagram shows part of the graph of a function with equation $y = f(x)$.

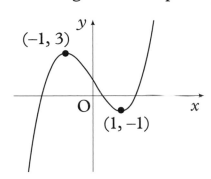

Which of the following diagrams shows the graph with equation

$y = -f(-x)$?

A
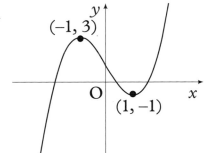

B
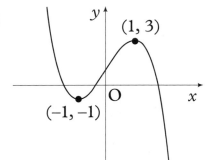

C
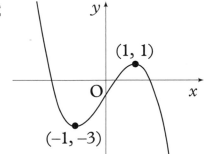

D
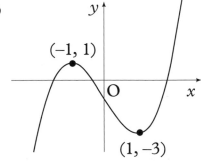

12. What is the solution of the equation $2\cos x - 1 = 0$ where $\dfrac{3\pi}{2} \leq x \leq 2\pi$?

A $\dfrac{\pi}{3}$

B $\dfrac{5\pi}{3}$

C $\dfrac{7\pi}{8}$

D $\dfrac{11\pi}{12}$

13. The expression $2x^2 + 8x + 3$ is re-written in the form $2(x + p)^2 + q$.

What is the value of q?

A -13

B -5

C 5

D 7

14. In the cuboid ABCD, EFGH, $\overrightarrow{AB} = \mathbf{x}$, $\overrightarrow{AD} = \mathbf{y}$ and $\overrightarrow{EA} = \mathbf{z}$.

Express $\overrightarrow{HB}$ in terms of $\mathbf{x}$, $\mathbf{y}$ and $\mathbf{z}$.

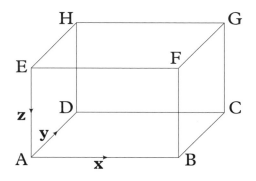

A $\overrightarrow{HB} = \mathbf{x} + \mathbf{y} + \mathbf{z}$

B $\overrightarrow{HB} = \mathbf{x} - \mathbf{y} + \mathbf{z}$

C $\overrightarrow{HB} = -\mathbf{x} + -\mathbf{y} - \mathbf{z}$

D $\overrightarrow{HB} = -\mathbf{x} + \mathbf{y} - \mathbf{z}$

15. Given that $f(x) = \cos 2x$, find $f'(x)$.

A $\sin 2x$

B $-\sin 2x$

C $2\sin 2x$

D $-2\sin 2x$

16. The vector $\overrightarrow{AB}$ is represented by $\begin{pmatrix} 2 \\ 4 \\ 6 \end{pmatrix}$ and B has coordinates $(8, 0, -2)$.

Find the coordinates of M, the midpoint of AB.

A $(9, 2, 1)$

B $(6, -4, -8)$

C $(5, 2, 2)$

D $(7, -2, -5)$

17. The diagram shows part of the graph whose equation is of the form $y = 6k^x$.

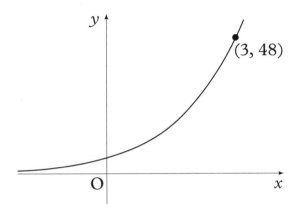

$(3, 48)$

What is the value of k?

A   64

B   16

C   6

D   2

18. Given that $\log_{10} y = 4\log_{10} x + \log_{10} 5$, express y in terms of x.

A $y = 4x + 5$

B $y = 20x$

C $y = 5x^4$

D $y = 5 \times 4^x$

19. In the following diagram, find the value of $\cos\angle DGF$.

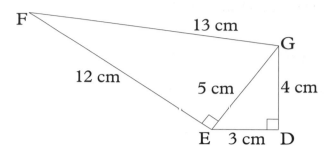

   A $-\dfrac{16}{65}$

   B $-\dfrac{4}{13}$

   C $\dfrac{16}{65}$

   D $\dfrac{56}{65}$

20. The product of the roots of the equation $(x+a)(2x+1)(x+3)=0$ is 6.
Find the value of a.

   A -4

   B   2

   C   3

   D   4

[END OF SECTION A]

SECTION B

ALL questions should be attempted

21. A function is defined on the set of real numbers by $f(x) = x^3 - 3x^2 + 4$.

(a) Find the coordinates of the stationary points on the curve $y = f(x)$ and determine their nature. **(6)**

(b) (i) Show that $(x + 1)$ is a factor of $x^3 - 3x^2 + 4$.

    (ii) Hence or otherwise factorise $x^3 - 3x^2 + 4$ fully. **(5)**

(c) State the coordinates of the points where the curve with equation $y = f(x)$ meets both the axes and hence sketch the curve. **(4)**

22. A curve has equation $y = \dfrac{12}{\sqrt{x}} + 2x$, where $x > 0$.

Find the equation of the tangent to the curve at the point where $x = 4$. **(6)**

23. Triangle KLM has vertices K (6, 2), L (−4, −4) and M (−1, 8).

(a) Find the equation of the median MN. **(3)**

(b) Find the equation of the altitude LP. **(3)**

(c) Find the coordinates of the point of intersection of MN and LP. **(3)**

[END OF SECTION B]

[END OF QUESTION PAPER]

Practice Paper C

PAPER ONE

Do not use a calculator

SECTION A

ALL questions should be attempted

1. What is the equation of the line joining the points $(2, 1)$ and $(3, 5)$?

   A $x - 4y + 2 = 0$

   B $4x - y - 2 = 0$

   C $x - 4y + 7 = 0$

   D $4x - y - 7 = 0$

2. A sequence is defined by the recurrence relation

   $u_{n+1} = 2u_n + 1$ with $u_0 = 3$.

   What is the value of u_2?

   A 7

   B 9

   C 10

   D 15

3. The point A has coordinates $(3, 4, 5)$, $\overrightarrow{AB}$ represents the vector $\begin{pmatrix} 3 \\ 3 \\ 3 \end{pmatrix}$ and

   $\overrightarrow{BC}$ represents the vector $\begin{pmatrix} 5 \\ 3 \\ 4 \end{pmatrix}$. What are the coordinates of C?

   A $(6, 7, 8)$

   B $(8, 7, 9)$

   C $(11, 10, 12)$

   D $(1, 4, 4)$

4. When $x^2 + 6x + 1$ is written in the form $(x + a)^2 + b$, what is the value of b?

    A   −10

    B   −8

    C   −5

    D   10

5. Given that $f(x) = 5\sqrt{x}$, find the value of $f'(4)$.

    A $\dfrac{5}{4}$

    B $\dfrac{5}{2}$

    C   5

    D $\dfrac{80}{3}$

6. A sequence is generated by the recurrence relation $u_{n+1} = \dfrac{1}{3}u_n + 11$ with $u_0 = -3$.
    What is the limit of this sequence as $n \to \infty$?

    A $\dfrac{1}{33}$

    B $\dfrac{33}{4}$

    C $\dfrac{33}{2}$

    D   33

7. The diagram shows the graph with equation of the form $y = p \sin qx$ for $0 \le x \le 2\pi$.

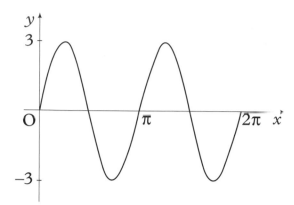

What is the equation of this graph?

A $y = 3 \sin 2x$

B $y = 3 \sin 3x$

C $y = 2 \sin 3x$

D $y = 6 \sin 2x$

8. What is the length of the radius of the circle $x^2 + y^2 - 6x - 8y + 7 = 0$?

A $3\sqrt{2}$

B $4\sqrt{2}$

C $\sqrt{93}$

D $\sqrt{107}$

9. The graph of the function $f(x) = ax^2 - 5x + 4$ does not cross or touch the x-axis.

What is the range of values of a?

A $a > \dfrac{25}{16}$

B $a \ge \dfrac{25}{16}$

C $a < \dfrac{25}{16}$

D $a \le \dfrac{25}{16}$

10. Find $\int\left(\sin x - \dfrac{3}{x^2}\right)dx$.

A $\quad -\cos x + \dfrac{3}{x} + C$

B $\quad \cos x + \dfrac{3}{x} + C$

C $\quad -\cos x + \dfrac{1}{x^3} + C$

D $\quad \cos x + \dfrac{1}{x^3} + C$

11. The vectors $\begin{pmatrix} 2 \\ -6 \\ 2 \end{pmatrix}$ and $\begin{pmatrix} 1 \\ -3 \\ m \end{pmatrix}$ are perpendicular.

Find the value of m.

A $\quad -10$

B $\quad -8$

C $\quad 0$

D $\quad 8$

12. Given that $f(x) = 2x - 1$ and $g(x) = x^2$, find $f(g(x))$.

A $\quad (2x - 1)x^2$

B $\quad x^2 + 2x - 1$

C $\quad 2x^2 - 1$

D $\quad (2x - 1)^2$

13. Which graph is most likely to be the graph of the function $f(x) = 2x^2 - 3x + 2$?

A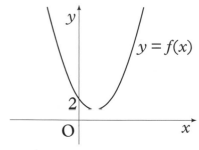

B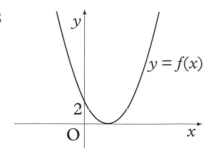

C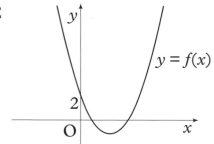

D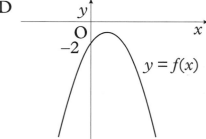

14. For which value of x does the function $f(x) = (x-3)(x+7)$ have a stationary value?

A   −7

B   −2

C   2

D   3

15. If $A(x) = 2x^2 - 5x + 7$, what is the rate of change of A with respect to x when $x = 3$?

A   3

B   7

C   10

D   28

16. What is the solution of the inequality $x^2 + 5x + 6 < 0$, where x is a real number?

A $-3 < x < -2$

B $x < -3, \ x > -2$

C $2 < x < 3$

D $x < 2, \ x > 3$

17. What is the remainder on dividing $(3x^3 - 4x + 8)$ by $(x - 2)$?

    A   −8

    B   12

    C   24

    D   28

18. Two equations involving k and a are

$$k \sin a° = \sqrt{3}$$

$$k \cos a° = -1$$

where $k > 0$ and $90 \leq a < 180$.

What are the values of k and a?

	k	a
A	2	150
B	2	120
C	$\sqrt{10}$	150
D	$\sqrt{10}$	120

19. The diagram shows graphs with equations $y = f(x)$ and $y = mx$. The coordinates of the points where the graphs cross each other and where the curve crosses the x-axis are as shown.

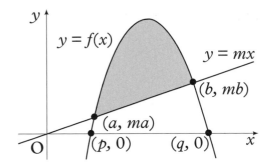

Which of the following represents the shaded area?

A $\displaystyle\int_{p}^{q} f(x)dx - \int_{a}^{b} mxdx$

B $\displaystyle\int_{p}^{q} [f(x)-mx]dx$

C $\displaystyle\int_{a}^{b} [f(x)-mx]dx$

D $\displaystyle\int_{ma}^{mb} [f(x)-mx]dx$

20. The diagram shows the graph of $y = f(x)$ where f is a logarithmic function.

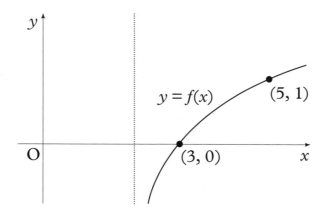

What is $f(x)$?

A $f(x) = \log_{5}(x-2)$

B $f(x) = \log_{3}(x-2)$

C $f(x) = \log_{3}(x-3)$

D $f(x) = \log_{5}(x-3)$

[END OF SECTION A]

SECTION B
ALL questions should be attempted

21. A triangle STU has vertices S (2, 4), T (0, −4) and U (8, 0).

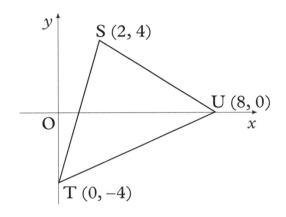

    (a) Find the equation of the line l_1, the median from U in triangle STU.   **(3)**

    (b) Find the equation of the line l_2, the perpendicular bisector of TU.   **(4)**

    (c) Find the coordinates of the point of intersection of the lines l_1 and l_2.   **(1)**

22. (a) Show that $(x-3)$ is a factor of $f(x) = 6x^3 - 25x^2 + x + 60$.   **(2)**

    (b) Hence factorise $f(x)$ fully.   **(3)**

23. The line with equation $5x + y - 15 = 0$ is a tangent at the point P to a circle with centre C (−3, 4).

    (a) Find the equation of the radius CP.   **(3)**

    (b) Find the coordinates of P.   **(3)**

    (c) What is the equation of the circle?   **(2)**

24. In the diagram, A is the point (3, 4) and the line OB has equation $y = \frac{1}{4}x$.

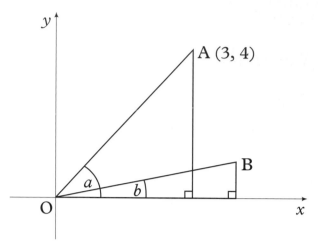

(a) (i) Explain why $\tan b = \frac{1}{4}$.

   (ii) Find the values of $\sin b$ and $\cos b$. **(3)**

(b) (i) Find the value of $\sin(a - b)$.

   (ii) Find the value of $\tan(a - b)$. **(6)**

[END OF SECTION B]

[END OF QUESTION PAPER]

Practice Paper D

PAPER ONE

Do not use a calculator

SECTION A

ALL questions should be attempted

1. A sequence is defined by $u_{n+1} = 0 \cdot 2u_n + 6$ with $u_5 = 20$.

   What is the value of u_7?

   A   6·4

   B   7·2

   C   7·4

   D   8

2. A circle has equation $2x^2 + 2y^2 + 12x - 8y - 30 = 0$.

   What are the coordinates of the centre of this circle?

   A $(6, -4)$

   B $(-6, 4)$

   C $(3, -2)$

   D $(-3, 2)$

3. Given that $\mathbf{u} = \begin{pmatrix} -4 \\ 3 \end{pmatrix}$ and $\mathbf{v} = \begin{pmatrix} -1 \\ -2 \end{pmatrix}$, what is the magnitude of the vector $(\mathbf{u} - \mathbf{v})$?

   A   8

   B $\sqrt{34}$

   C $\sqrt{29}$

   D   2

4. Given that $(x - 1)$ is a factor of $(x^3 - 5x^2 + kx - 5)$, find the value of k.

   A   11

   B   9

   C   −1

   D   −11

5. What is the equation of the straight line passing through the point (2, 3) and with gradient -4?

    A $4x + y - 14 = 0$

    B $4x + y - 11 = 0$

    C $4x + y + 5 = 0$

    D $4x + y + 10 = 0$

6. The minimum value of $2\sin(x - 60)° - 8$ is a.

    This minimum value occurs when $x = b$.

    What are the values of a and b?

	a	b
A	-8	150
B	-10	330
C	-8	60
D	-10	240

7. The diagram shows a circle, centre (2, 4) and a tangent drawn at the point (5, 8).

    What is the equation of this tangent?

    A $y - 8 = -\dfrac{4}{3}(x - 5)$

    B $y - 8 = -\dfrac{3}{4}(x - 5)$

    C $y - 8 = \dfrac{3}{4}(x - 5)$

    D $y - 8 = \dfrac{4}{3}(x - 5)$

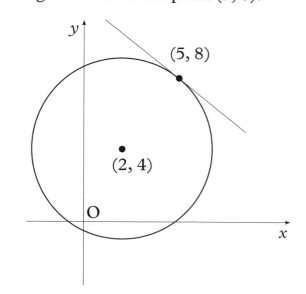

8. Given that $f(x) = x^2 - 9$ and $g(x) = \sqrt{x}$, find $f(g(x))$.

   A $\quad x - 3$

   B $\quad x - 9$

   C $\quad \sqrt{x-3}$

   D $\quad \sqrt{x^2 - 9}$

9. Find $\int (2x-1)^3 \, dx$.

   A $\quad \dfrac{1}{4}(2x-1)^4 + C$

   B $\quad \dfrac{1}{2}(2x-1)^4 + C$

   C $\quad \dfrac{1}{8}(2x-1)^4 + C$

   D $\quad (x^2 - x)^4 + C$

10. PQRS is a parallelogram. T is a point on RS such that RT : TS = 1 : 2.

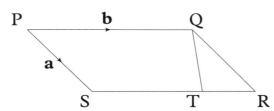

   Given that $\overrightarrow{PS}$ and $\overrightarrow{PQ}$ represent the vectors **a** and **b** respectively, which of the following represents $\overrightarrow{QT}$?

   A $\quad \mathbf{a} - \dfrac{1}{3}\mathbf{b}$

   B $\quad \mathbf{a} - \dfrac{1}{2}\mathbf{b}$

   C $\quad \mathbf{a} + \dfrac{1}{3}\mathbf{b}$

   D $\quad \mathbf{a} + \dfrac{2}{3}\mathbf{b}$

11. The points $(3, 6, 4)$, $(5, 3, 1)$ and $(9, a, b)$ are collinear.

    What are the values of a and b?

    A   7 and 5

    B   8 and 6

    C   0 and -2

    D -3 and -5

12. Given that $f(x) = \sin 2x$, evaluate $f'\left(\dfrac{\pi}{6}\right)$.

    A $\dfrac{\sqrt{3}}{2}$

    B $\dfrac{1}{2}$

    C   1

    D $\sqrt{3}$

13. A function is given by $f(x) = \sqrt{x^2 - 4}$.

    What is a suitable domain of f?

    A $x \geq 2$

    B $x \leq 2$

    C $-2 \leq x \leq 2$

    D $x \leq -2,\ x \geq 2$

14. The expression $3x^2 + 18x + 1$ is re-written in the form $3(x + a)^2 + b$.

    What is the value of b?

    A -80

    B -26

    C -8

    D   82

15. Which of the following sketches shows the position of the graph of the function
$f(x) = 2x^2 - 4x + 2$?

A
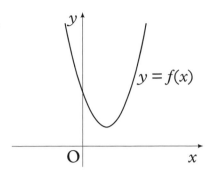

B
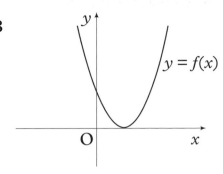

C
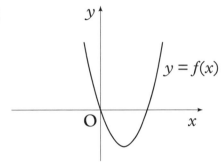

D
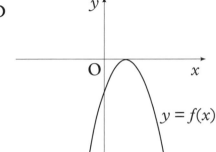

16. Which of the following could be part of the graph of a function f such that
$f'(x) = x(x-1)$?

A
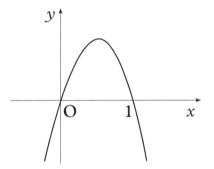

B
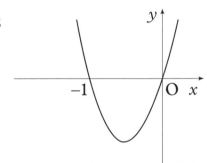

C
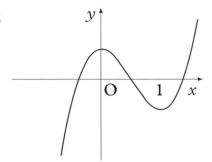

D
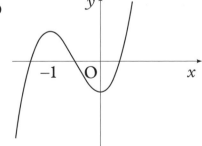

17. Vectors **a** and **b** are inclined at 60° to each other.

If $|\mathbf{a}| = 6$ and $|\mathbf{b}| = 5$, what is the value of **a.b**?

A   15

B $15\sqrt{3}$

C $5\dfrac{1}{2}$

D $30\sqrt{3}$

18. In triangle ABC, angle $C = 60°$. What is the value of $\sin A \cos B + \cos A \sin B$?

A $-\dfrac{\sqrt{3}}{2}$

B $-\dfrac{1}{2}$

C $\dfrac{1}{2}$

D $\dfrac{\sqrt{3}}{2}$

19. The diagram shows the graph of $y = f(x)$ where f is a logarithmic function.

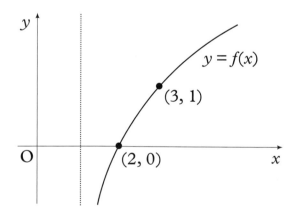

What is $f(x)$?

A $f(x) = \log_4(x-1)$

B $f(x) = \log_2(x+1)$

C $f(x) = \log_2(x-1)$

D $f(x) = \log_4(x+1)$

20. Given that $\log_{10} y = 4\log_{10} x + \log_{10} 2$, express y in terms of x.

A $\quad y = 8x$

B $\quad y = 4x + 2$

C $\quad y = x^4 + 2$

D $\quad y = 2x^4$

[END OF SECTION A]

SECTION B

ALL questions should be attempted

21. The points A and B have coordinates $(-7, -2)$ and $(11, 4)$ respectively.

    (a) Find the coordinates of the point P which divides AB in the ratio 2 : 1. **(2)**

    (b) Show that the equation of the line l through P perpendicular to AB is $y = -3x + 17$. **(3)**

    (c) C is the point $(-3, -4)$.

    The line through C parallel to AB meets l at Q.

    Show that Q is the point $(6, -1)$. **(3)**

22. A function is defined by the formula $f(x) = x^3 - 12x$.

    (a) Find where the graph of $y = f(x)$ meets the x and y-axes. **(2)**

    (b) Find the coordinates of the stationary points of the function and determine their nature. **(7)**

    (c) Sketch the graph of $y = f(x)$. **(2)**

23. The equation of a circle is $x^2 + y^2 - 8x - 4y = 0$.

    (a) State the coordinates of the centre and calculate the radius of the circle. **(2)**

    (b) Find the coordinates of the points of intersection, A and B, of the line $x + y = 8$ and the given circle. **(5)**

    (c) Write down the coordinates of M, the midpoint of AB. **(1)**

    (d) A second circle has the same centre as the given circle and has AB as a tangent. What is the equation of this second circle? **(3)**

[END OF SECTION B]

[END OF QUESTION PAPER]

Answer Support for Paper One (Non-calculator) A–D

Practice Paper A

PAPER ONE

SECTION A

1. Use List of Formulae:

   The equation $x^2 + y^2 + 2gx + 2fy + c = 0$ represents a circle with centre $(-g, -f)$ and radius $\sqrt{g^2 + f^2 - c}$.

   Hence radius $= \sqrt{g^2 + f^2 - c} = \sqrt{5^2 + (-2)^2 - (-35)} = \sqrt{25 + 4 + 35} = \sqrt{64} = 8$.

   Correct answer is **D**.

2. $f(x) = 2x^3 - 4 \Rightarrow f'(x) = 6x^2 \Rightarrow f'(2) = 6 \times (2)^2 = 6 \times 4 = 24$.

   Correct answer is **B**.

3. Gradient of line, $m = \dfrac{y_2 - y_1}{x_2 - x_1} = \dfrac{12 - 0}{0 - 6} = \dfrac{12}{-6} = -2$.

   Hence equation of line is $y = -2x + 12$ (using the formula $y = mx + c$).

   When $x = -2$, $y = -2 \times (-2) + 12 = 16 \Rightarrow a = 16$.

   Correct answer is **A**.

4. $u_{n+1} = 0 \cdot 1 u_n + 28 \Rightarrow u_2 = 0 \cdot 1 u_1 + 28 = 0 \cdot 1 \times 20 + 28 = 2 + 28 = 30$.

   $u_3 = 0 \cdot 1 u_2 + 28 = 0 \cdot 1 \times 30 + 28 = 3 + 28 = 31$.

   $u_4 = 0 \cdot 1 u_3 + 28 = 0 \cdot 1 \times 31 + 28 = 3 \cdot 1 + 28 = 31 \cdot 1$.

   Correct answer is **D**.

5. Use List of Formulae: $\sin 2A = 2\sin A \cos A$.

   To use this formula, we must find $\cos x$ when $\sin x = \dfrac{1}{3}$.

   Use Pythagoras' Theorem:

   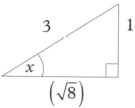

   $\sin 2x = 2\sin x \cos x = 2 \times \dfrac{1}{3} \times \dfrac{\sqrt{8}}{3} = \dfrac{2\sqrt{8}}{9} = \dfrac{2\sqrt{4 \times 2}}{9} = \dfrac{4\sqrt{2}}{9}$. $\left(\text{Simplify } \sqrt{8}\right)$

   Correct answer is **D**.

6. $3x^2 + \dfrac{1}{x} = 3x^2 + x^{-1}$.

   Derivative is $6x - x^{-2} = 6x - \dfrac{1}{x^2}$.

   Correct answer is **B**.

7. Re-arrange $x - 5y + 9 = 0$ into the form $y = mx + c$.

   $x - 5y + 9 = 0 \Rightarrow 5y = x + 9 \Rightarrow y = \dfrac{1}{5}x + \dfrac{9}{5} \Rightarrow m = \dfrac{1}{5}$.

   Hence the perpendicular line gradient is -5.

   Correct answer is **D**.

8. Use List of Formulae:

   The equation $(x - a)^2 + (y - b)^2 = r^2$ represents a circle with centre (a, b) and radius r.

   Centre is midpoint of AB $= \left(\dfrac{2 + 12}{2}, \dfrac{4 + 4}{2}\right) = (7, 4)$.

   Since AB is a horizontal line, distance AB = 10 units, hence radius = 5 units.

   Hence equation is $(x - 7)^2 + (y - 4)^2 = 25$.

   Correct answer is **C**.

9. Use the discriminant to find roots of $7x^2 + 4x + 1 = 0$.

$b^2 - 4ac = 4^2 - 4 \times 7 \times 1 = 16 - 28 = -12$.

Since $b^2 - 4ac < 0$, there are no real roots.

Correct answer is **D**.

10. Choose two values of x to substitute in $f(x) = \log_3 x$.

Appropriate values are $x = 1$ and $x = 3$.

When $x = 1$, $f(1) = \log_3 1 = 0$, and when $x = 3$, $f(3) = \log_3 3 = 1$.

Hence graph passes through the points $(1, 0)$ and $(3, 1)$.

Correct answer is **A**.

11. To solve this quadratic inequality, sketch the graph of $y = x^2 - 2x - 15$.

$x^2 - 2x - 15 = (x + 3)(x - 5)$, hence the graph cuts the x-axis at -3 and 5.

The $(+)1$ in front of x^2 indicates a minimum turning point.

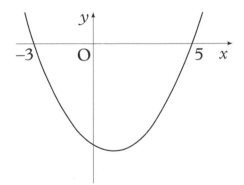

The function is positive when the graph is above the x-axis.

Hence the solution is $x < -3$, $x > 5$.

Correct answer is **D**.

12. The limit is found using the formula $L = \dfrac{b}{1-a} = \dfrac{12}{1-0\cdot4} = \dfrac{12}{0\cdot6} = \dfrac{120}{6} = 20$.

Correct answer is **A**.

13. To find the maximum or minimum value of $f(x) = a\sin x° + b\cos x°$, it should be expressed in the form $k\cos(x-\alpha)°$. The maximum value is then k, and the minimum value $-k$. For $f(x) = 4\sin x° - 3\cos x°$, $k = \sqrt{4^2 + (-3)^2} = 5$.

Hence the minimum value is -5.

Correct answer is **C**.

14. You may prefer to convert $\dfrac{7\pi}{6}$ to degrees; $\dfrac{7\pi}{6} = (180 \div 6 \times 7)° = 210°$.

Hence $\sin\dfrac{7\pi}{6} = \sin 210° = -\sin 30° = -\dfrac{1}{2}$.

Correct answer is **B**.

15. $\overrightarrow{AB} = \mathbf{b} - \mathbf{a} = \begin{pmatrix} 2 \\ 0 \\ 8 \end{pmatrix} - \begin{pmatrix} 3 \\ -5 \\ 7 \end{pmatrix} = \begin{pmatrix} -1 \\ 5 \\ 1 \end{pmatrix}$.

$AB = \sqrt{(-1)^2 + 5^2 + 1^2} = \sqrt{1 + 25 + 1} = \sqrt{27}$.

Both statements are correct.

Correct answer is **D**.

16. $y = 2x^3 - 4x \Rightarrow \dfrac{dy}{dx} = 6x^2 - 4 = 6 \times (-2)^2 - 4 = 6 \times 4 - 4 = 20$ when $x = -2$.

Hence the gradient is 20.

Correct answer is **C**.

17. Use List of Formulae: $\cos(A - B) = \cos A \cos B + \sin A \sin B$.

$\cos 4x° \cos x° + \sin 4x° \sin x° = \dfrac{1}{2} \Rightarrow \cos(4x - x)° = \dfrac{1}{2} \Rightarrow \cos 3x° = \dfrac{1}{2}$.

For the smallest positive value of x, $3x = 60 \Rightarrow x = 20$.

Correct answer is **B**.

18. Remember that the gradient of a line is the tangent of the angle between the line and the positive direction of the x-axis ($m = \tan\theta$).

By calculation, angle FEX = $\dfrac{5\pi}{6}$ radians or 150°.

Hence the gradient of EF = $\tan 150° = -\tan 30° = -\dfrac{1}{\sqrt{3}}$.

Correct answer is **A**.

19. $f(x) = \sqrt{(7 - 5x^2)} = (7 - 5x^2)^{\frac{1}{2}} \Rightarrow f'(x) = \dfrac{1}{2}(7 - 5x^2)^{-\frac{1}{2}} \times (-10x)$ (using chain rule).

$f'(x) = \dfrac{1}{2}(7 - 5x^2)^{-\frac{1}{2}} \times (-10x) = \dfrac{-10x}{2(7 - 5x^2)^{\frac{1}{2}}} = \dfrac{-5x}{\sqrt{(7 - 5x^2)}}$.

Correct answer is **A**.

20. Shaded area = $\displaystyle\int_{-2}^{2} \left[(4 - x^2) - (x^2 - 4)\right] dx = \int_{-2}^{2} (8 - 2x^2) dx$.

Correct answer is **B**.

SECTION B

21. To draw the graph of $y = 2 - f(x)$, think of it as $y = -f(x) + 2$, then reflect the graph of $y = f(x)$ in the x-axis, followed by moving it up 2 units in the direction of the y-axis e.g. $(3, 5) \rightarrow (3, -5) \rightarrow (3, -3)$.

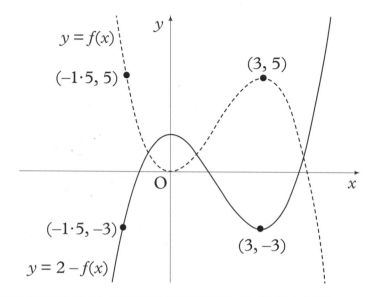

3 marks

22. NOTE: Although results obtained from an accurate drawing will receive **no marks**, it is a good idea to make an accurate drawing in order to help you organise your working and keep you on the right track.

(a) $M = \left(\dfrac{4+5}{2}, \dfrac{7+(-2)}{2} \right) = \left(4\frac{1}{2}, 2\frac{1}{2} \right)$.

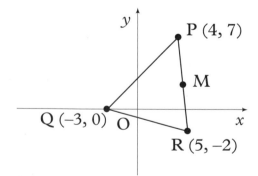

(b) Gradient of QM $= \dfrac{y_2 - y_1}{x_2 - x_1} = \dfrac{2\frac{1}{2} - 0}{4\frac{1}{2} + 3} = \dfrac{2\frac{1}{2}}{7\frac{1}{2}} = \dfrac{5}{15} = \dfrac{1}{3}$.

Equation of QM is $y - b = m(x - a) \Rightarrow y - 0 = \dfrac{1}{3}(x + 3) \Rightarrow 3y = x + 3$.

(c) Gradient of PQ $= \dfrac{y_2 - y_1}{x_2 - x_1} = \dfrac{0 - 7}{-3 - 4} = \dfrac{-7}{-7} = 1$.

So gradient of altitude RT $= -1$.

Equation of RT is $y - b = m(x - a) \Rightarrow y + 2 = -1(x - 5) \Rightarrow y + 2 = -x + 5$.

This simplifies to $y = -x + 3$.

(d) RT meets QM at T, so solve simultaneous equations to find T.

Solve:

$$3y = x + 3 \qquad (1)$$
$$y = -x + 3 \qquad (2)$$

(1) + (2): $4y = 6$

Hence $y = 1\frac{1}{2}$

and, by substitution, $x = 1\frac{1}{2}$.

T is the point $(1\frac{1}{2}, 1\frac{1}{2})$.

9 marks

23. $y = x^3 - 3x^2 - 24x + 5 \Rightarrow \dfrac{dy}{dx} = 3x^2 - 6x - 24 = 0$ at stationary points.

$3x^2 - 6x - 24 = 0 \Rightarrow 3(x^2 - 2x - 8) = 0 \Rightarrow 3(x+2)(x-4) = 0.$

Hence stationary points occur when $x = -2$ or 4.

When $x = -2$, $y = (-2)^3 - 3 \times (-2)^2 - 24 \times (-2) + 5 = -8 - 12 + 48 + 5 = 33.$

When $x = 4$, $y = (4)^3 - 3 \times (4)^2 - 24 \times (4) + 5 = 64 - 48 - 96 + 5 = -75.$

Nature table:

x	$\rightarrow$	-2	$\rightarrow$	4	$\rightarrow$
$\dfrac{dy}{dx}$	$+$	0	$-$	0	$+$
	$\nearrow$	max	$\searrow$	min	$\nearrow$

Hence $(-2, 33)$ is a maximum turning point and $(4, -75)$ is a minimum turning point.

8 marks

24. (a) A difficult trigonometric identity! Use List of Formulae:

$\cos(A + B) = \cos A \cos B - \sin A \sin B$

$\sin 2A = 2 \sin A \cos A$

$\cos 2A = 2\cos^2 A - 1$

$$
\begin{aligned}
\text{Left side} &= \cos 3x \\
&= \cos(2x + x) \\
&= \cos 2x \cos x - \sin 2x \sin x \\
&= (2\cos^2 x - 1)\cos x - (2\sin x \cos x)\sin x \\
&= 2\cos^3 x - \cos x - 2\sin^2 x \cos x \\
&= 2\cos^3 x - \cos x - 2(1 - \cos^2 x)\cos x \quad (\text{from } \cos^2 x + \sin^2 x = 1) \\
&= 2\cos^3 x - \cos x - 2\cos x + 2\cos^3 x \\
&= 4\cos^3 x - 3\cos x \\
&= \text{Right side.}
\end{aligned}
$$

NOTE: This is very demanding and the majority of students will struggle here. However, as the answer is given, you can use it for parts (b) and (c).

(b) $\cos 3x = 4\cos^3 x - 3\cos x \Rightarrow 4\cos^3 x = 3\cos x + \cos 3x$

$$\Rightarrow \cos^3 x = \frac{1}{4}(3\cos x + \cos 3x).$$

(c) Use List of Formulae for integration: $\int \cos ax = \frac{1}{a}\sin ax.$

$$\text{Area} = \int_0^{\frac{\pi}{2}} \cos^3 x\, dx.$$

$$\text{So } \int_0^{\frac{\pi}{2}} \cos^3 x\, dx = \frac{1}{4}\int_0^{\frac{\pi}{2}} (3\cos x + \cos 3x)\, dx$$

$$= \frac{1}{4}\left[3\sin x + \frac{1}{3}\sin 3x \right]_0^{\frac{\pi}{2}}$$

$$= \frac{1}{4}\left[\left(3\sin\frac{\pi}{2} + \frac{1}{3}\sin\frac{3\pi}{2} \right) - \left(3\sin 0 + \frac{1}{3}\sin 0 \right) \right]$$

$$= \frac{1}{4}\left(3 - \frac{1}{3} - 0 \right)$$

$$= \frac{1}{4} \times \frac{8}{3}$$

$$= \frac{2}{3}.$$

10 marks

Answer Support for Paper One (Non-calculator) A–D

Practice Paper B

PAPER ONE

SECTION A

1. $u_{n+1} = 0 \cdot 4u_n + 8 \Rightarrow u_8 = 0 \cdot 4u_7 + 8 = 0 \cdot 4 \times 20 + 8 = 8 + 8 = 16$.

   $u_9 = 0 \cdot 4u_8 + 8 = 0 \cdot 4 \times 16 + 8 = 6 \cdot 4 + 8 = 14 \cdot 4$.

   Correct answer is **D**.

2. Use List of Formulae:

   The equation $x^2 + y^2 + 2gx + 2fy + c = 0$ represents a circle with centre $(-g, -f)$ and radius $\sqrt{g^2 + f^2 - c}$.

   Hence centre $= (-g, -f) = (-4, 5)$.

   Correct answer is **C**.

3. $3y = 2x + 12$ meets the x-axis where $y = 0 \Rightarrow 0 = 2x + 12 \Rightarrow x = -6$.

   Hence P is the point $(-6, 0)$. Q is the point $(3, -5)$. So the gradient of PQ is

   $$\frac{y_2 - y_1}{x_2 - x_1} = \frac{-5 - 0}{3 - (-6)} = \frac{-5}{9} = -\frac{5}{9}.$$

   Correct answer is **B**.

4. The simplest method is to consider the x-coordinates of the three collinear points.

   $-5 \to -1 \to 7$. The jump from -5 to -1 is 4, and the jump from -1 to 7 is 8.

   Hence Q divides PR in the ratio $4 : 8 = 1 : 2$.

   The correct answer is **B**.

5. $\int (2x+1)^2\, dx = \dfrac{1}{3\times 2}(2x+1)^3 + C = \dfrac{1}{6}(2x+1)^3 + C.$

   Correct answer is **C**.

6. $f(g(-1)) = f\left\{(-1)^2\right\} = f(1) = 1 - 2\times 1 = 1 - 2 = -1.$

   Correct answer is **D**.

7. Use List of Formulae:

   The equation $(x-a)^2 + (y-b)^2 = r^2$ represents a circle with centre (a, b) and radius r.

   Here the centre is $(-5, -8)$ and, by inspection of diagram, the radius is 8.

   Therefore the equation is $(x+5)^2 + (y+8)^2 = 64$.

   Correct answer is **D**.

8. $\dfrac{d+10}{2} = 4 \Rightarrow d+10 = 8 \Rightarrow d = -2;\ \dfrac{8+e}{2} = -2 \Rightarrow 8+e = -4 \Rightarrow e = -12.$

   Correct answer is **A**.

9. The limit is found using the formula $L = \dfrac{b}{1-a} = \dfrac{-80}{1-0\cdot 8} = \dfrac{-80}{0\cdot 2} = \dfrac{-800}{2} = -400.$

   Correct answer is **A**.

10. $b^2 - 4ac = (2p)^2 - 4 \times 1 \times p^2 = 4p^2 - 4p^2 = 0.$

Hence the roots are real and equal, so both statements are correct.

Correct answer is **D**.

11. Given the graph of $y = f(x)$, you draw the graph of $y = -f(-x)$ by giving $f(x)$ a half turn rotation about the origin, i.e. $(-1, 3) \to (1, -3)$ and $(1, -1) \to (-1, 1)$.

(Alternatively, first reflect the graph in the y-axis to find $f(-x)$ then reflect this in the x-axis to find $-f(-x)$.)

Correct answer is **D**.

12. $2\cos x - 1 = 0 \Rightarrow \cos x = \dfrac{1}{2} \Rightarrow x = \dfrac{\pi}{3}, \dfrac{5\pi}{3}$.

In the interval $\dfrac{3\pi}{2} \leq x \leq 2\pi$, only $\dfrac{5\pi}{3}$ is suitable.

NOTE: You may prefer to convert from radians to degrees leading to $300°$, and then convert back to radians.

Correct answer is **B**.

13. $2x^2 + 8x + 3 = 2(x^2 + 4x) + 3 = 2(x^2 + 4x + 4) + 3 - 8 = 2(x + 2)^2 - 5 \Rightarrow q = -5.$

NOTE: This technique is known as completing the square.

Correct answer is **B**.

14. $\overrightarrow{HB} = \overrightarrow{HE} + \overrightarrow{EA} + \overrightarrow{AB} = -\mathbf{y} + \mathbf{z} + \mathbf{x} = \mathbf{x} - \mathbf{y} + \mathbf{z}.$

Correct answer is **B**.

15. Use List of Formulae:

    If $f(x) = \cos ax$, $f'(x) = -a \sin ax$.

    Hence if $f(x) = \cos 2x$, $f'(x) = -2 \sin 2x$.

    Correct answer is **D**.

16. $\overrightarrow{AB} = \mathbf{b} - \mathbf{a} \Rightarrow \begin{pmatrix} 8 \\ 0 \\ -2 \end{pmatrix} - \mathbf{a} = \begin{pmatrix} 2 \\ 4 \\ 6 \end{pmatrix} \Rightarrow \mathbf{a} = \begin{pmatrix} 8 \\ 0 \\ -2 \end{pmatrix} - \begin{pmatrix} 2 \\ 4 \\ 6 \end{pmatrix} = \begin{pmatrix} 6 \\ -4 \\ -8 \end{pmatrix}$.

    Therefore A is the point $(6, -4, -8)$. As B is $(8, 0, -2)$, the coordinates of M, the

    midpoint of AB, are $\left(\dfrac{6+8}{2}, \dfrac{-4+0}{2}, \dfrac{-8+(-2)}{2} \right) = (7, -2, -5)$.

    Correct answer is **D**.

17. Substitute $(3, 48)$ into the equation $y = 6k^x$.

    Hence $48 = 6k^3 \Rightarrow k^3 = \dfrac{48}{6} = 8 \Rightarrow k = 2$.

    Correct answer is **D**.

18. $\log_{10} y = 4\log_{10} x + \log_{10} 5 = \log_{10} x^4 + \log_{10} 5 = \log_{10} 5x^4 \Rightarrow y = 5x^4$.

    Correct answer is **C**.

19. Use List of Formulae: $\cos(A+B) = \cos A \cos B - \sin A \sin B$.

    $\cos \angle DGF = \cos(\angle DGE + \angle EGF) = \cos \angle DGE \cos \angle EGF - \sin \angle DGE \sin \angle EGF$

    $= \dfrac{4}{5} \times \dfrac{5}{13} - \dfrac{3}{5} \times \dfrac{12}{13} = \dfrac{20}{65} - \dfrac{36}{65} = -\dfrac{16}{65}$.

    Correct answer is **A**.

20. The roots of the equation $(x+a)(2x+1)(x+3)=0$ are $-a$, $-\dfrac{1}{2}$ and -3.

The product of the roots is $-a \times \left(-\dfrac{1}{2}\right) \times (-3) = 6 \Rightarrow a = -4$.

The correct answer is **A**.

SECTION B

21. (a) $f(x)=x^3-3x^2+4 \Rightarrow f'(x)=3x^2-6x=0$ at stationary points.

$3x^2-6x=0 \Rightarrow 3x(x-2)=0 \Rightarrow x=0$ or 2.

Hence stationary points occur when $x=0$ or 2.

When $x=0$, $f(x)=4$.

When $x=2$, $f(x)=2^3-3\times(2)^2+4=8-12+4=0$.

Nature table:

x	$\rightarrow$	0	$\rightarrow$	2	$\rightarrow$
$f'(x)$	+	0	−	0	+
	/	‾	\	_	/
		max		min	

Hence $(0, 4)$ is a maximum turning point and

$(2, 0)$ is a minimum turning point.

(b) (i)

$$\begin{array}{c|cccc} -1 & 1 & -3 & 0 & 4 \\ & & -1 & 4 & -4 \\ \hline & 1 & -4 & 4 & 0 \end{array}$$

The remainder is 0 so $x+1$ is a factor.

(ii) $x^3-3x^2+4=(x+1)(x^2-4x+4)=(x+1)(x-2)(x-2)$.

(c) $y = f(x) = x^3 - 3x^2 + 4$ meets the y-axis when $x = 0$, i.e. at $(0, 4)$.

$y = f(x)$ meets the x-axis when $y = 0$, i.e. when $(x + 1)(x - 2)(x - 2) = 0$, i.e. at $(-1, 0)$ and $(2, 0)$.

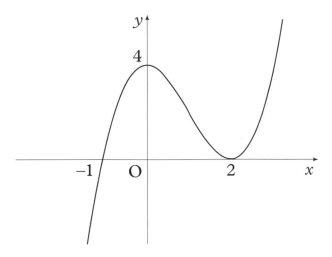

15 marks

22. $y = \dfrac{12}{\sqrt{x}} + 2x = \dfrac{12}{x^{\frac{1}{2}}} + 2x = 12x^{-\frac{1}{2}} + 2x \Rightarrow \dfrac{dy}{dx} = -6x^{-\frac{3}{2}} + 2 = \dfrac{-6}{x^{\frac{3}{2}}} + 2 = \dfrac{-6}{\sqrt{x^3}} + 2$.

The gradient of the tangent $= \dfrac{-6}{\sqrt{x^3}} + 2 = \dfrac{-6}{\sqrt{4^3}} + 2 = -\dfrac{6}{8} + 2 = \dfrac{5}{4}$ when $x = 4$.

When $x = 4$, $y = \dfrac{12}{\sqrt{x}} + 2x = \dfrac{12}{\sqrt{4}} + 2 \times 4 = 14$. So point of contact is $(4, 14)$.

Equation of tangent is $y - b = m(x - a) \Rightarrow y - 14 = \dfrac{5}{4}(x - 4)$.

This simplifies to $4y = 5x + 36$.

6 marks

23. (a) The median MN joins M to N, the midpoint of KL.

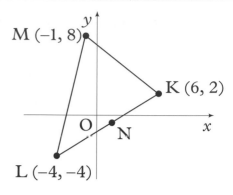

As K is (6, 2) and L is (−4, −4), N is $\left(\dfrac{6+(-4)}{2}, \dfrac{2+(-4)}{2}\right) = (1, -1)$.

The gradient of MN is $\dfrac{y_2 - y_1}{x_2 - x_1} = \dfrac{-1-8}{1+1} = \dfrac{-9}{2} = -\dfrac{9}{2}$.

The equation of MN is $y - b = m(x - a) \Rightarrow y + 1 = -\dfrac{9}{2}(x - 1)$.

This simplifies thus: $y + 1 = -\dfrac{9}{2}(x - 1) \Rightarrow 2y + 2 = -9x + 9 \Rightarrow 9x + 2y = 7$.

(b) The altitude LP is perpendicular to KM.

The gradient of KM is $\dfrac{y_2 - y_1}{x_2 - x_1} = \dfrac{8-2}{-1-6} = \dfrac{6}{-7} = -\dfrac{6}{7}$.

Therefore the gradient of LP is $\dfrac{7}{6}$.

The equation of LP is $y - b = m(x - a) \Rightarrow y + 4 = \dfrac{7}{6}(x + 4)$.

This simplifies thus: $y + 4 = \dfrac{7}{6}(x + 4) \Rightarrow 6y + 24 = 7x + 28 \Rightarrow 7x - 6y = -4$.

(c) To find the point of intersection of MN and LP, solve simultaneous equations.

$$9x + 2y = 7 \qquad (1)$$

$$7x - 6y = -4 \qquad (2)$$

$(1) \times 3: \qquad 27x + 6y = 21 \qquad (3)$

$(2) + (3): \qquad 34x = 17$

$$\Rightarrow x = \frac{1}{2}.$$

By substitution (1): $9 \times \dfrac{1}{2} + 2y = 7 \Rightarrow 2y = \dfrac{5}{2} \Rightarrow y = \dfrac{5}{4}.$

Hence point of intersection is $\left(\frac{1}{2}, 1\frac{1}{4} \right)$.

9 marks

Answer Support for Paper One (Non-calculator) A–D

Practice Paper C

PAPER ONE

SECTION A

1. Gradient of line $= \dfrac{y_2 - y_1}{x_2 - x_1} = \dfrac{5-1}{3-2} = \dfrac{4}{1} = 4$.

   Equation of line is $y - b = m(x - a) \Rightarrow y - 1 = 4(x - 2) \Rightarrow y - 1 = 4x - 8$.

   This simplifies to $4x - y - 7 = 0$.

   Correct answer is **D**.

2. $u_{n+1} = 2u_n + 1 \Rightarrow u_1 = 2u_0 + 1 = 2 \times 3 + 1 = 7$.

   $u_2 = 2u_1 + 1 = 2 \times 7 + 1 = 15$.

   Correct answer is **D**.

3. $\overrightarrow{AC} = \overrightarrow{AB} + \overrightarrow{BC} = \begin{pmatrix} 3 \\ 3 \\ 3 \end{pmatrix} + \begin{pmatrix} 5 \\ 3 \\ 4 \end{pmatrix} = \begin{pmatrix} 8 \\ 6 \\ 7 \end{pmatrix}$.

   $\overrightarrow{AC} = \mathbf{c} - \mathbf{a} = \mathbf{c} - \begin{pmatrix} 3 \\ 4 \\ 5 \end{pmatrix} = \begin{pmatrix} 8 \\ 6 \\ 7 \end{pmatrix} \Rightarrow \mathbf{c} = \begin{pmatrix} 11 \\ 10 \\ 12 \end{pmatrix}$. Hence C is (11, 10, 12).

   The correct answer is **C**.

4. $x^2 + 6x + 1 = x^2 + 6x + 9 + 1 - 9 = (x+3)^2 - 8 \Rightarrow b = -8.$

Correct answer is **B**.

5. $f(x) = 5\sqrt{x} = 5x^{\frac{1}{2}} \Rightarrow f'(x) = \frac{5}{2}x^{-\frac{1}{2}} = \frac{5}{2x^{\frac{1}{2}}} = \frac{5}{2\sqrt{x}} \Rightarrow f'(4) = \frac{5}{2 \times \sqrt{4}} = \frac{5}{4}$

Correct answer is **A**.

6. The limit is found using the formula $L = \dfrac{b}{1-a} = \dfrac{11}{1-\dfrac{1}{3}} = \dfrac{11}{\dfrac{2}{3}} = 11 \times \dfrac{3}{2} = \dfrac{33}{2}.$

Correct answer is **C**.

7. As amplitude is 3, and there are 2 cycles between 0 and 2π, the equation is

$y = 3\sin 2x.$

Correct answer is **A**.

8. Use List of Formulae:

The equation $x^2 + y^2 + 2gx + 2fy + c = 0$ represents a circle with centre $(-g, -f)$
and radius $\sqrt{g^2 + f^2 - c}$.

Hence radius $= \sqrt{g^2 + f^2 - c} = \sqrt{(-3)^2 + (-4)^2 - 7} = \sqrt{9 + 16 - 7} = \sqrt{18}$.

$\sqrt{18} = \sqrt{9 \times 2} = 3\sqrt{2}.$

Correct answer is **A**.

9. If $f(x) = ax^2 - 5x + 4$ does not cut or touch the x-axis, then the equation

$ax^2 - 5x + 4 = 0$ has no real roots and therefore $b^2 - 4ac < 0$.

Hence $b^2 - 4ac = (-5)^2 - 4 \times a \times 4 < 0 \Rightarrow 25 - 16a < 0 \Rightarrow 16a > 25 \Rightarrow a > \dfrac{25}{16}$.

Correct answer is **A**.

10. Use List of Formulae for integration: $\displaystyle\int \sin ax = -\frac{1}{a}\cos ax + C$.

$\displaystyle\int \left(\sin x - \frac{3}{x^2} \right) dx = \int \left(\sin x - 3x^{-2} \right) dx = -\cos x + 3x^{-1} + C$.

This simplifies to $-\cos x + \dfrac{3}{x} + C$.

Correct answer is **A**.

11. The vectors $\begin{pmatrix} 2 \\ -6 \\ 2 \end{pmatrix}$ and $\begin{pmatrix} 1 \\ -3 \\ m \end{pmatrix}$ are perpendicular if $\begin{pmatrix} 2 \\ -6 \\ 2 \end{pmatrix} \cdot \begin{pmatrix} 1 \\ -3 \\ m \end{pmatrix} = 0$.

Use List of Formulae: $\mathbf{a.b} = a_1b_1 + a_2b_2 + a_3b_3$ where $\mathbf{a} = \begin{pmatrix} a_1 \\ a_2 \\ a_3 \end{pmatrix}$ and $\mathbf{b} = \begin{pmatrix} b_1 \\ b_2 \\ b_3 \end{pmatrix}$.

Hence $2 \times 1 + (-6) \times (-3) + 2 \times m = 0$

$\Rightarrow 2 + 18 + 2m = 0 \Rightarrow 2m = -20 \Rightarrow m = -10$.

Correct answer is **A**.

12. $f(g(x)) = f(x^2) = 2x^2 - 1$.

Correct answer is **C**.

13. Find the nature of the roots of $2x^2 - 3x + 2 = 0$.

$b^2 - 4ac = (-3)^2 - 4 \times 2 \times 2 = 9 - 16 = -7$. So no real roots.

Hence the parabola does not cut or touch the x-axis.

The $(+)2$ in front of x^2 indicates a minimum turning point.

Correct answer is **A**.

14. A function $f(x)$ has a stationary value when $f'(x) = 0$.

$f(x) = (x - 3)(x + 7) = x^2 + 7x - 3x - 21 = x^2 + 4x - 21 \Rightarrow f'(x) = 2x + 4$.

Hence stationary value occurs when $2x + 4 = 0 \Rightarrow x = -2$.

(This answer can also be found by considering the roots of the equation $(x - 3)(x + 7) = 0$. These are 3 and -7. The axis of symmetry of the parabola lies midway between them at -2. This is where the stationary value occurs.)

Correct answer is **B**.

15. Questions on rate of change are solved by differentiation.

$A(x) = 2x^2 - 5x + 7 \Rightarrow A'(x) = 4x - 5 = 4 \times 3 - 5 = 7$ when $x = 3$.

Correct answer is **B**.

16. To solve this quadratic inequality, sketch the graph of $y = x^2 + 5x + 6$.

$x^2 + 5x + 6 = (x + 2)(x + 3)$, hence the graph cuts the x-axis at -3 and -2.

The $(+)1$ in front of x^2 indicates a minimum turning point.

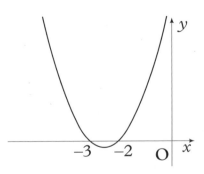

The function is negative when the graph is below the x-axis.

Hence the solution is $-3 < x < -2$.

Correct answer is **A**.

17.

$$2 \begin{array}{|rrrr} 3 & 0 & -4 & 8 \\ & 6 & 12 & 16 \\ \hline 3 & 6 & 8 & 24 \end{array}$$

Hence remainder is 24.

Correct answer is **C**.

---

18. $k = \sqrt{(\sqrt{3})^2 + (-1)^2} = 2; \ \tan a° = -\dfrac{\sqrt{3}}{1} \Rightarrow a = 120.$

The correct answer is **B**.

---

19. Area between two curves is $\displaystyle\int_{a}^{b} [f(x) - g(x)]\,dx.$

In this case, shaded area $= \displaystyle\int_{a}^{b} [f(x) - mx]\,dx.$

Correct answer is **C**.

---

20. Remember that $\log_a 1 = 0$ and $\log_a a = 1$ for any base.

The correct equation is of the form $f(x) = \log_a(x - b).$

Substitute $(3, 0)$ into this equation $\Rightarrow 0 = \log_a(3 - b).$

As $\log_a 1 = 0$, $b = 2.$

Now substitute $(5, 1)$ into $f(x) = \log_a(x - 2) \Rightarrow 1 = \log_a(5 - 2) = \log_a 3.$

As $\log_a a = 1$, $a = 3.$ Hence $f(x) = \log_3(x - 2).$

Correct answer is **B**.

21. (a) The median l_1 joins U to the midpoint of ST.

As S is (2, 4) and T is (0, −4), the midpoint is $\left(\dfrac{2+0}{2}, \dfrac{4+(-4)}{2}\right) = (1, 0)$.

The gradient of l_1 is $\dfrac{y_2 - y_1}{x_2 - x_1} = \dfrac{0-0}{1-8} = 0$.

The equation of l_1 is $y - b = m(x-a) \Rightarrow y - 0 = 0(x-8) \Rightarrow y = 0$.

(b) The midpoint of TU is $\left(\dfrac{0+8}{2}, \dfrac{-4+0}{2}\right) = (4, -2)$.

The gradient of TU is $\dfrac{y_2 - y_1}{x_2 - x_1} = \dfrac{0+4}{8-0} = \dfrac{4}{8} = \dfrac{1}{2}$.

The gradient of the perpendicular bisector l_2 is therefore -2.

The equation of l_2 is $y - b = m(x-a) \Rightarrow y + 2 = -2(x-4)$.

This simplifies to $y = -2x + 6$.

(c) Lines l_1 and l_2 intersect where $y = 0$ meets $y = -2x + 6$.

This occurs at the point (3, 0).

8 marks

22. (a)
$$
\begin{array}{r|rrrr}
3 & 6 & -25 & 1 & 60 \\
  &   & 18 & -21 & -60 \\
\hline
  & 6 & -7 & -20 & 0
\end{array}
$$

As remainder is 0, $(x-3)$ is a factor of $f(x) = 6x^3 - 25x^2 + x + 60$.

(b) $f(x) = 6x^3 - 25x^2 + x + 60 = (x-3)(6x^2 - 7x - 20)$

$\qquad\qquad = (x-3)(3x+4)(2x-5)$.

5 marks

23. (a) $5x + y - 15 = 0 \Rightarrow y = -5x + 15$.

Hence the tangent at P has gradient -5.

Therefore the radius CP at P has gradient $\dfrac{1}{5}$.

Equation of CP is $y - b = m(x - a) \Rightarrow y - 4 = \dfrac{1}{5}(x + 3) \Rightarrow 5y - 20 = x + 3$.

This simplifies to $5y = x + 23$.

(b) P is at the intersection of lines $y = -5x + 15$ and $5y = x + 23$.

Solve simultaneous equations:

$$y = -5x + 15 \Rightarrow y + 5x = 15 \qquad (1)$$
$$5y = x + 23 \Rightarrow 5y - x = 23 \qquad (2)$$
$$(2) \times 5: \qquad 25y - 5x = 115 \qquad (3)$$
$$(1) + (3): \qquad 26y = 130$$
$$y = 5$$

By substitution, when $y = 5$, $x = 2$, so P is the point $(2, 5)$.

(c) Use distance formula to find radius CP:

$$d = \sqrt{(x_2 - x_1)^2 + (y_2 - y_1)^2} = \sqrt{(2 + 3)^2 + (5 - 4)^2} = \sqrt{5^2 + 1^2} = \sqrt{26}.$$

Use List of Formulae to find equation of circle:

The equation $(x - a)^2 + (y - b)^2 = r^2$ represents a circle with centre (a, b) and radius r.

The equation is $(x + 3)^2 + (y - 4)^2 = 26$.

8 marks

24. (a) (i) $y = \frac{1}{4}x \Rightarrow m = \frac{1}{4}$ for line OB, and since $m = \tan\theta$, $\tan b = \frac{1}{4}$.

(ii) $\sqrt{4^2 + 1^2} = \sqrt{17} \Rightarrow \sin b = \frac{1}{\sqrt{17}}$ and $\cos b = \frac{4}{\sqrt{17}}$.

(b) Use List of Formulae: $\sin(A - B) = \sin A \cos B - \cos A \sin B$

$$\cos(A - B) = \cos A \cos B + \sin A \sin B$$

(i) $OA = \sqrt{3^2 + 4^2} = \sqrt{25} = 5 \Rightarrow \sin a = \frac{4}{5}$ and $\cos a = \frac{3}{5}$.

$$\sin(a - b) = \sin a \cos b - \cos a \sin b$$

$$= \frac{4}{5} \times \frac{4}{\sqrt{17}} - \frac{3}{5} \times \frac{1}{\sqrt{17}} = \frac{16}{5\sqrt{17}} - \frac{3}{5\sqrt{17}} = \frac{13}{5\sqrt{17}}.$$

(ii) $\cos(a - b) = \cos a \cos b + \sin a \sin b$

$$= \frac{3}{5} \times \frac{4}{\sqrt{17}} + \frac{4}{5} \times \frac{1}{\sqrt{17}} = \frac{12}{5\sqrt{17}} + \frac{4}{5\sqrt{17}} = \frac{16}{5\sqrt{17}}.$$

$$\tan(a - b) = \frac{\sin(a - b)}{\cos(a - b)} = \left(\frac{13}{5\sqrt{17}}\right) \div \left(\frac{16}{5\sqrt{17}}\right) = \frac{13}{16}.$$

9 marks

Answer Support for Paper One (Non-calculator) A–D

Practice Paper D

PAPER ONE

SECTION A

1. $u_{n+1} = 0 \cdot 2u_n + 6 \Rightarrow u_6 = 0 \cdot 2u_5 + 6 = 0 \cdot 2 \times 20 + 6 = 4 + 6 = 10$.

$u_7 = 0 \cdot 2u_6 + 6 = 0 \cdot 2 \times 10 + 6 = 2 + 6 = 8$.

Correct answer is **D**.

2. Use List of Formulae:

The equation $x^2 + y^2 + 2gx + 2fy + c = 0$ represents a circle with centre $(-g, -f)$.

$2x^2 + 2y^2 + 12x - 8y - 30 = 0 \Rightarrow x^2 + y^2 + 6x - 4y - 15 = 0$.

Hence centre is $(-3, 2)$.

Correct answer is **D**.

3. $\mathbf{u} - \mathbf{v} = \begin{pmatrix} -4 \\ 3 \end{pmatrix} - \begin{pmatrix} -1 \\ -2 \end{pmatrix} = \begin{pmatrix} -3 \\ 5 \end{pmatrix} \Rightarrow \left| \begin{pmatrix} -3 \\ 5 \end{pmatrix} \right| = \sqrt{(-3)^2 + 5^2} = \sqrt{9 + 25} = \sqrt{34}$.

Correct answer is **B**.

4.

$$\begin{array}{c|cccc} 1 & 1 & -5 & k & -5 \\ & & 1 & -4 & k-4 \\ \hline & 1 & -4 & (k-4) & 0 \end{array}$$

Hence $-5 + k - 4 = 0 \Rightarrow k = 9$.

Correct answer is **B**.

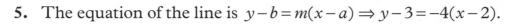

5. The equation of the line is $y - b = m(x - a) \Rightarrow y - 3 = -4(x - 2)$.

This simplifies to $y - 3 = -4x + 8 \Rightarrow 4x + y - 11 = 0$.

Correct answer is **B**.

6. The minimum value of $2\sin(x - 60)° - 8$ occurs when $\sin(x - 60)° = -1$.

Hence the minimum value is $2 \times (-1) - 8 = -2 - 8 = -10$.

The minimum value occurs when $\sin(x - 60)° = -1 \Rightarrow x - 60 = 270 \Rightarrow x = 330$.

Hence $a = -10$ and $b = 330$.

Correct answer is **B**.

7. The gradient of the radius is $\dfrac{y_2 - y_1}{x_2 - x_1} = \dfrac{8 - 4}{5 - 2} = \dfrac{4}{3}$.

Hence the gradient of the tangent is $-\dfrac{3}{4}$.

The equation of the tangent is $y - b = m(x - a) \Rightarrow y - 8 = -\dfrac{3}{4}(x - 5)$.

Correct answer is **B**.

8. $f(g(x)) = f\left(\sqrt{x}\right) = \left(\sqrt{x}\right)^2 - 9 = x - 9$.

Correct answer is **B**.

9. $\displaystyle\int (2x - 1)^3\, dx = \dfrac{1}{4 \times 2}(2x - 1)^4 + C = \dfrac{1}{8}(2x - 1)^4 + C$.

Correct answer is **C**.

10. If T is a point on RS such that RT : TS = 1 : 2, then $\overrightarrow{ST} = \frac{2}{3}\overrightarrow{SR}$.

$$\overrightarrow{QT} = \overrightarrow{QP} + \overrightarrow{PS} + \overrightarrow{ST} = \overrightarrow{QP} + \overrightarrow{PS} + \frac{2}{3}\overrightarrow{SR} = -\mathbf{b} + \mathbf{a} + \frac{2}{3}\mathbf{b} = \mathbf{a} - \frac{1}{3}\mathbf{b}.$$

Alternatively, you could use the pathway $\overrightarrow{QT} = \overrightarrow{QR} + \overrightarrow{RT}$.

Correct answer is **A**.

11. $\begin{pmatrix} 5 \\ 3 \\ 1 \end{pmatrix} - \begin{pmatrix} 3 \\ 6 \\ 4 \end{pmatrix} = \begin{pmatrix} 2 \\ -3 \\ -3 \end{pmatrix}$, and $\begin{pmatrix} 9 \\ a \\ b \end{pmatrix} - \begin{pmatrix} 5 \\ 3 \\ 1 \end{pmatrix} = \begin{pmatrix} 4 \\ a-3 \\ b-1 \end{pmatrix} = \begin{pmatrix} 4 \\ -6 \\ -6 \end{pmatrix}$ as this must be a multiple of

$\begin{pmatrix} 2 \\ -3 \\ -3 \end{pmatrix}$ if the points are collinear.

Hence $a - 3 = -6 \Rightarrow a = -3$, and $b - 1 = -6 \Rightarrow b = -5$.

Correct answer is **D**.

12. Use List of Formulae:

If $f(x) = \sin ax$, $f'(x) = a\cos ax$.

Hence if $f(x) = \sin 2x$, $f'(x) = 2\cos 2x$.

Hence $f'\left(\frac{\pi}{6}\right) = 2\cos\left(2 \times \frac{\pi}{6}\right) = 2\cos\left(\frac{\pi}{3}\right) = 2 \times \frac{1}{2} = 1$.

Correct answer is **C**.

13. As the square root of a negative number does not exist, the domain is restricted to $x^2 - 4 \geq 0$.

To solve this quadratic inequality, sketch the graph of $y = x^2 - 4$.

$x^2 - 4 = (x+2)(x-2)$, hence the graph cuts the x-axis at -2 and 2.

The $(+)1$ in front of x^2 indicates a minimum turning point.

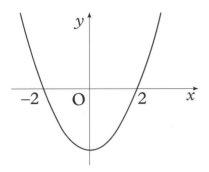

The function is positive when the graph is above the x-axis.

Hence the solution is $x \leq -2$, $x \geq 2$.

Correct answer is **D**.

14. $3x^2 + 18x + 1 = 3(x^2 + 6x) + 1 = 3(x^2 + 6x + 9) + 1 - 27 = 3(x+3)^2 - 26$.

Hence $b = -26$.

NOTE: This technique is known as completing the square.

Correct answer is **B**.

15. As $b^2 - 4ac = (-4)^2 - 4 \times 2 \times 2 = 16 - 16 = 0$, there are equal roots.

So the graph touches the x-axis.

The $(+)2$ in front of x^2 indicates a minimum turning point.

Correct answer is **B**.

16. A function such that $f'(x) = x(x-1)$ has stationary points when $f'(x) = 0$.

Therefore there are stationary points when $x(x-1) = 0$, i.e. when $x = 0$ or 1.

By inspecting the graphs, graph C satisfies these conditions.

Correct answer is **C**.

17. Use List of Formulae: $\mathbf{a}.\mathbf{b} = |\mathbf{a}||\mathbf{b}|\cos\theta$ where θ is the angle between $\mathbf{a}$ and $\mathbf{b}$.

$\mathbf{a}.\mathbf{b} = |\mathbf{a}||\mathbf{b}|\cos 60° = 6 \times 5 \times \dfrac{1}{2} = 15$.

Correct answer is **A**.

18. Use List of Formulae: $\sin(A+B) = \sin A \cos B + \cos A \sin B$.

A + B + C = 180°, so as C = 60°, then A + B = 120°.

$\sin(A+B) = \sin A \cos B + \cos A \sin B \Rightarrow \sin A \cos B + \cos A \sin B = \sin 120°$.

$\sin 120° = \sin 60° = \dfrac{\sqrt{3}}{2}$.

Correct answer is **D**.

19. Remember that $\log_a 1 = 0$ and $\log_a a = 1$ for any base.

The correct equation is of the form $f(x) = \log_a(x+b)$.

Substitute (2, 0) into this equation $\Rightarrow 0 = \log_a(2+b)$. As $\log_a 1 = 0$, $b = -1$.

Now substitute (3, 1) into $f(x) = \log_a(x-1) \Rightarrow 1 = \log_a(3-1) = \log_a 2$.

As $\log_a a = 1$, $a = 2$. Hence $f(x) = \log_2(x-1)$.

Correct answer is **C**.

20. $\log_{10} y = 4\log_{10} x + \log_{10} 2 = \log_{10} x^4 + \log_{10} 2 = \log_{10} 2x^4 \Rightarrow y = 2x^4$.

Correct answer is **D**.

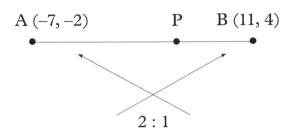

21. (a) Use the section formula.

A $(-7, -2)$ P   B $(11, 4)$

2 : 1

Hence P is $\left(\dfrac{1\times(-7)+2\times 11}{3}, \dfrac{1\times(-2)+2\times 4}{3} \right) = (5,\ 2)$.

(Alternatively x_A to x_B is (-7) to 11 ($= 18$).

Point P lies $\dfrac{2}{3}$ of the way, so $x_P = -7 + \dfrac{2}{3}(18)$

$$= -7 + 12$$
$$= 5.)$$

(b) The gradient of AB is $\dfrac{y_2 - y_1}{x_2 - x_1} = \dfrac{4+2}{11+7} = \dfrac{6}{18} = \dfrac{1}{3}$.

The gradient of line l is -3.

The equation of line l is $y - b = m(x - a) \Rightarrow y - 2 = -3(x - 5)$.

This simplifies to $y - 2 = -3x + 15 \Rightarrow y = -3x + 17$.

(c) The line through C parallel to AB has gradient $\dfrac{1}{3}$.

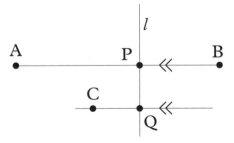

The equation of this line is $y - b = m(x - a) \Rightarrow y + 4 = \dfrac{1}{3}(x + 3)$.

This simplifies to $y+4=\dfrac{1}{3}(x+3)\Rightarrow 3y+12=x+3\Rightarrow 3y=x-9$.

Solve simultaneous equations to find where it meets l.

$$3y = x - 9 \qquad (1)$$
$$y = -3x + 17 \qquad (2)$$

$(1)\times 3: \qquad 9y = 3x - 27 \qquad (3)$

$(2)+(3): 10y = -10$

$$\Rightarrow y = -1$$

Substitute in (1): $3\times(-1)=x-9\Rightarrow x=-3+9=6$.

Therefore Q is the point $(6, -1)$.

(This could also have been done by substituting the coordinates of Q into equations (1), (2).)

8 marks

---

22. (a) $y = f(x)$ meets the x-axis where $y=0\Rightarrow x^3-12x=0\Rightarrow x(x^2-12)=0$.

This occurs where $x=0$ or $x^2=12\Rightarrow x=\pm\sqrt{12}$.

$y=f(x)$ meets the y-axis where $x=0\Rightarrow y=0^3-12\times 0=0$.

Hence the graph meets the axes at $(-\sqrt{12},0)$, $(0,0)$ and $(\sqrt{12},0)$.

(b) $y=x^3-12x\Rightarrow \dfrac{dy}{dx}=3x^2-12=0$ at stationary points.

$3x^2-12=0\Rightarrow 3(x^2-4)=0\Rightarrow 3(x+2)(x-2)=0$.

Hence stationary points occur when $x=-2$ or 2.

When $x=-2$, $y=(-2)^3-12\times(-2)=-8+24=16$.

When $x=2$, $y=(2)^3-12\times(2)=8-24=-16$.

Nature table:

x	$\rightarrow$	-2	$\rightarrow$	2	$\rightarrow$
$\dfrac{dy}{dx}$	$+$	0	$-$	0	$+$
	$\diagup$	$\overline{}$ max	$\diagdown$	$\underline{}$ min	$\diagup$

Hence $(-2, 16)$ is a maximum turning point and

$(2, -16)$ is a minimum turning point.

(c)

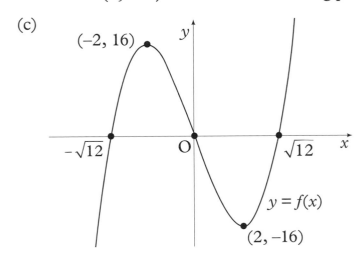

$(-2, 16)$

$-\sqrt{12}$

$\sqrt{12}$

$y = f(x)$

$(2, -16)$

11 marks

23. (a) Use List of Formulae:

The equation $x^2 + y^2 + 2gx + 2fy + c = 0$ represents a circle with centre

$(-g, -f)$ and radius $\sqrt{g^2 + f^2 - c}$.

Hence centre is $(4, 2)$ and radius is $\sqrt{(-4)^2 + (-2)^2 - 0} = \sqrt{16 + 4} = \sqrt{20} = 2\sqrt{5}$.

(b) $x + y = 8 \Rightarrow y = 8 - x$. Substitute into equation of circle:

$x^2 + (8 - x)^2 - 8x - 4(8 - x) = 0$

$\Rightarrow x^2 + 64 - 16x + x^2 - 8x - 32 + 4x = 0$

$\Rightarrow 2x^2 - 20x + 32 = 0$

$\Rightarrow x^2 - 10x + 16 = 0$

$\Rightarrow (x - 2)(x - 8) = 0$

$\Rightarrow x = 2$ or 8

When $x = 2$, $y = 6$ and when $x = 8$, $y = 0$ (using $y = 8 - x$).

Hence A is $(2, 6)$ and B $(8, 0)$.

(c) M, the midpoint of AB, is $\left(\dfrac{2+8}{2}, \dfrac{6+0}{2}\right) = (5, 3)$.

(d) Use List of Formulae:

The equation $(x - a)^2 + (y - b)^2 = r^2$ represents a circle with centre (a, b) and radius r.

By using symmetry, M is the point of contact of the tangent.

Use distance formula to find radius:

$$r = \sqrt{(x_2 - x_1)^2 + (y_2 - y_1)^2} = \sqrt{(5-4)^2 + (3-2)^2} = \sqrt{1^2 + 1^2} = \sqrt{2}$$

Hence equation of circle is:

$$(x - a)^2 + (y - b)^2 = r^2 \Rightarrow (x - 4)^2 + (y - 2)^2 = 2.$$

11 marks

Introduction to Paper Two

In this section, there are four complete Paper Two question papers. They are modelled as closely as possible on recent SQA exam papers.

In Paper Two, you **are** allowed to use a calculator.

In Paper Two, ALL questions should be attempted. In total, Paper Two is worth 60 marks. The time allocated to complete Paper Two is 1 hour 10 minutes.

There is a set of solutions for each practice paper. These will include some hints and advice on how to tackle particular questions.

In all papers, you are allowed to use a list of formulae. You will find this list on the following page.

Each SQA Paper Two begins with the following instructions:

Read carefully

1 **Calculators may be used in this paper.**

2 Full credit will be given only where the solution contains appropriate working.

3 Answers obtained by readings from scale drawings will not receive any credit.

LIST OF FORMULAE

Circle:

The equation $x^2 + y^2 + 2gx + 2fy + c = 0$ represents a circle with centre $(-g, -f)$ and radius $\sqrt{g^2 + f^2 - c}$.

The equation $(x - a)^2 + (y - b)^2 = r^2$ represents a circle with centre (a, b) and radius r.

Scalar Product: $\mathbf{a}.\mathbf{b} = |\mathbf{a}||\mathbf{b}|\cos\theta$, where θ is the angle between $\mathbf{a}$ and $\mathbf{b}$

or $\qquad \mathbf{a}.\mathbf{b} = a_1b_1 + a_2b_2 + a_3b_3$ where $\mathbf{a} = \begin{pmatrix} a_1 \\ a_2 \\ a_3 \end{pmatrix}$ and $\mathbf{b} = \begin{pmatrix} b_1 \\ b_2 \\ b_3 \end{pmatrix}$.

Trigonometric formulae:

$$\sin(A \pm B) = \sin A \cos B \pm \cos A \sin B$$
$$\cos(A \pm B) = \cos A \cos B \mp \sin A \sin B$$
$$\sin 2A = 2\sin A \cos A$$
$$\cos 2A = \cos^2 A - \sin^2 A$$
$$= 2\cos^2 A - 1$$
$$= 1 - 2\sin^2 A$$

Table of standard derivatives:

$f(x)$	$f'(x)$
$\sin ax$	$a\cos ax$
$\cos ax$	$-a\sin ax$

Table of standard integrals:

$f(x)$	$\int f(x)dx$
$\sin ax$	$-\dfrac{1}{a}\cos ax + C$
$\cos ax$	$\dfrac{1}{a}\sin ax + C$

Practice Paper A

PAPER TWO

Calculators may be used in this paper

ALL questions should be attempted

1. A, B and C are the points $(1, -6, 2)$, $(3, 0, -2)$ and $(6, 9, -8)$ respectively.

   (a) Show that A, B and C are collinear. **(3)**

   (b) Find the ratio in which B divides AC. **(1)**

2. Functions f and g are given by $f(x) = 2x - 3$ and $g(x) = x^2 + 2$.

   (a) (i) Find $h(x)$ where $h(x) = f(g(x))$.

       (ii) Find $k(x)$ where $k(x) = g(f(x))$. **(3)**

   (b) Find the value of a such that $k(a) = 2h(a) - 15$. **(2)**

3. Solve, for x, the equation $2\cos 2x° - 4\cos x° - 1 = 0$, where $0 \leq x \leq 360$. **(6)**

4. (a) Prove that the function $f(x) = x^3 + 9x^2 + 27x$ is never decreasing. **(5)**

   (b) Evaluate $\displaystyle\int_{1}^{8}\left(\frac{x-2}{\sqrt[3]{x}}\right)dx$. **(5)**

5. Given that $3x - 1$ is a factor of $6x^3 - 17x^2 + kx + 12$, find the value of k and then factorise the expression fully when k has this value. **(5)**

6. Vectors **a**, **b** and **c** are shown in the following diagram and angle PSR $= 60°$.

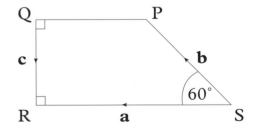

   Given that $|\mathbf{a}| = 6$ and $|\mathbf{b}| = 5$

   (a) Evaluate $\mathbf{a} \cdot (\mathbf{b} + \mathbf{c})$. **(3)**

   (b) Find $|\mathbf{b} + \mathbf{c}|$. **(2)**

7. (a) Express $8\cos x^\circ - 15\sin x^\circ$ in the form $k\cos(x+\alpha)^\circ$, where $k > 0$ and $0 \leq \alpha \leq 90$. **(4)**

   (b) Hence find algebraically the values of x between 0 and 360 for which $8\cos x^\circ - 15\sin x^\circ = 12$. **(4)**

   (c) Find the range of values of x between 0 and 360 for which $8\cos x^\circ - 15\sin x^\circ \geq 12$. **(2)**

8. A liquid cools according to the law $T_t = T_0 e^{-kt}$, where $T_0(°C)$ is the initial temperature and T_t (°C) is the temperature after t minutes.

   (a) If a liquid cools from 100°C to 80°C in 2 minutes, calculate the value of k. **(3)**

   (b) How long will it take for the liquid to cool to half its original temperature? **(3)**

9. (a) Find the coordinates of the centre, C, and the radius of the circle whose equation is $x^2 + y^2 + 14x + 4y + 19 = 0$. **(2)**

   (b) The point A (−12, 1) lies on the circle. Find the equation of the tangent to the circle at A. **(3)**

   (c) Prove that the point B (−9, 6) lies on this tangent. **(1)**

   (d) Find the equation of the circle which passes through the points C, A and B. **(3)**

[END OF QUESTION PAPER]

Practice Paper B

PAPER TWO

Calculators may be used in this paper

ALL questions should be attempted

1. A and B are the points (4, 4, 10) and (10, −2, 4) respectively.

   The point C lies between A and B and divides AB in the ratio 5 : 1.

   (a) Find the coordinates of C. **(2)**

   (b) If D is the point $(x, -2, 2)$ and CD is perpendicular to AB, find x. **(5)**

2. A circle has equation $x^2 + y^2 - 6x - 10y + 26 = 0$.

   (a) State the coordinates of C, the centre of the circle. **(1)**

   (b) Find the equation of the chord with midpoint M (4, 6). **(3)**

   (c) Show that the line with equation $x + y = 12$ is a tangent to the circle and find the coordinates of P, the point of contact. **(6)**

3. Solve the equation $4\sin 2x° - 6\cos x° = 0$, where $0 \le x \le 360$. **(5)**

4. (a) Express $\cos\theta - \sqrt{3}\sin\theta$ in the form $k\cos(\theta - \alpha)$ where $k > 0$ and $0 < \alpha < 2\pi$. **(4)**

   (b) Hence find the derivative of $\cos\theta - \sqrt{3}\sin\theta$ as a single trigonometric expression. **(2)**

5. A rectangular garden is laid out as shown in the diagram with a rectangular lawn of area 72 square metres surrounded by a border.

   The length of the lawn is x metres.

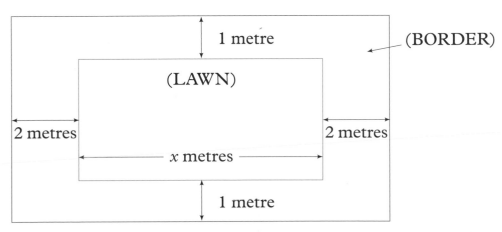

(a) If the total area of the garden is given as A square metres, show that

$$A = 80 + 2x + \frac{288}{x}.$$ (3)

(b) Find the dimensions of the garden with minimum area.

Justify your answer. (6)

6. Functions $f(x) = \sin x$, $g(x) = \cos x$, $h(x) = x - \frac{\pi}{3}$ and $k(x) = x - \frac{\pi}{6}$ are defined on a suitable set of real numbers.

(a) Find expressions for

   (i) $f(h(x))$.

   (ii) $g(k(x))$. (3)

(b) (i) Show that $f(h(x)) = \frac{1}{2}\sin x - \frac{\sqrt{3}}{2}\cos x$.

   (ii) Find a similar expression for $g(k(x))$ and hence solve the equation

$$f(h(x)) - g(k(x)) = \frac{\sqrt{3}}{2} \text{ for } 0 \le x \le 2\pi.$$ (6)

7. (a) If $\log_{12} x + \log_{12} y = 1$ for $x, y > 0$, show that $x = \frac{12}{y}$. (2)

(b) Hence, or otherwise, solve the equations

$$\log_{12} x + \log_{12} y = 1$$
$$\text{and } 4x + y = 19.$$ (4)

8. The parabola shown in the diagram has equation $y = x^2 - 9$.

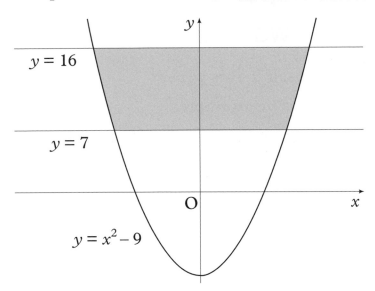

$y = 16$

$y = 7$

$y = x^2 - 9$

The shaded area lies between the lines $y = 7$ and $y = 16$.

Calculate the shaded area.　　　　　　　　　　　　　　　**(8)**

[END OF QUESTION PAPER]

Practice Paper C

PAPER TWO

Calculators may be used in this paper
ALL questions should be attempted

1. (a) Show that the line with equation $y - 5x + 24 = 0$ is a tangent to the parabola with equation $y = x^2 - 3x - 8$. **(4)**

   (b) Find the coordinates of the point of contact. **(1)**

2. D, F and E are the points $(2, -1, 1)$, $(6, -3, 5)$ and $(8, -4, 7)$ respectively.

   (a) Show that D, F and E are collinear and find the ratio in which F divides DE. **(4)**

   (b) P is the point $(a, 2, 4)$ and DP is perpendicular to EP.

   Calculate the value of a. **(5)**

   (c) Show that triangle DPF is isosceles and calculate the size of angle DPF. **(5)**

3. Solve, for x, the equation $\cos 2x° - 4 \sin x° + 5 = 0$, where $0 \leq x \leq 360$. **(6)**

4. Show that the tangent to the curve $y = \sqrt{x} + \dfrac{6}{\sqrt{x}}$ at the point $(4, 5)$ is

   perpendicular to the line with equation $8x - y = 5$. **(5)**

5. (a) The expression $5 \cos x° - 12 \sin x°$ may be written in the form $k \cos(x + \alpha)°$, where $k > 0$ and $0 \leq \alpha \leq 360$.

   Calculate the values of k and α. **(4)**

   (b) Find the maximum and minimum values of the expression $5 \cos x° - 12 \sin x°$.

   Determine the values of x, in the interval $0 \leq x \leq 360$, at which these maximum and minimum values occur. **(3)**

   (c) Find where the graph of $y = 5 \cos x° - 12 \sin x°$ crosses the y-axis and hence sketch the graph of $y = 5 \cos x° - 12 \sin x°$, where $0 \leq x \leq 360$. **(3)**

6. The curve in the following diagram has equation $y = x^3 - x + 3$.

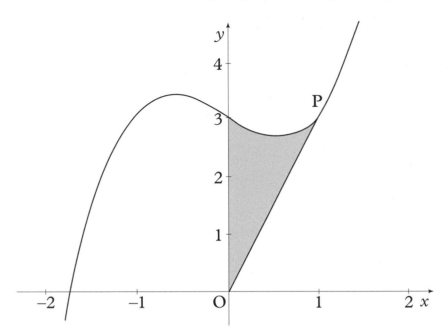

Show that the point P $(1, 3)$ lies on this curve, and calculate the area bounded by the curve, the line OP and the y-axis, as shaded in the diagram. **(6)**

7. The shaded rectangle in the following diagram is drawn with one vertex at the origin and the opposite vertex at E on the line CD where C is the point $(0, 6)$ and D is the point $(18, 0)$.

The other two vertices of the rectangle are the points P $(0, p)$ and Q $(q, 0)$.

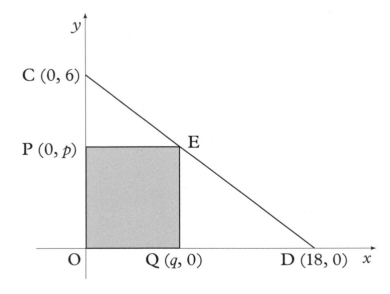

(a) Show that $p = \frac{1}{3}(18 - q)$ and hence deduce that the area, $A(q)$, of the rectangle is given by $A(q) = 6q - \frac{1}{3}q^2$. (3)

(b) Find the value of q which produces the largest area of the rectangle. (5)

8. (a) Show that $\log_{16} x = \frac{1}{2}\log_4 x$. (2)

(b) If, for $x > 1$, $\log_{16} x = (\log_4 x)^2$, find the exact value of x. (4)

[END OF QUESTION PAPER]

Practice Paper D

PAPER TWO

Calculators may be used in this paper

ALL questions should be attempted

1. The diagram shows a cuboid OABC, DEFG.

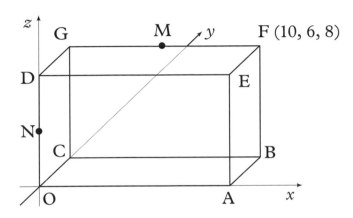

   F is the point (10, 6, 8).

   M is the midpoint of GF.

   N is the midpoint of OD.

   (a) State the coordinates of M. **(1)**

   (b) Write down the components of $\overrightarrow{MF}$ and $\overrightarrow{MN}$. **(2)**

   (c) Find the size of angle FMN. **(5)**

2. Given that $f(x) = \sin 2x + \cos^2 x$, where x is defined on the set of real numbers, find the equation of the tangent to the curve with equation $y = f(x)$ at the point where $x = \dfrac{\pi}{4}$. **(7)**

3. A hospital patient is to be given 75 milligrams of a painkiller every 6 hours.

   In any 6-hour period, the amount of painkiller in the body reduces by 20%.

   (a) The painkiller becomes effective when there is a continuous minimum of 150 milligrams in the body. Calculate how many doses will be necessary before the painkiller becomes effective. **(3)**

   (b) Write down a recurrence relation for the amount of painkiller in the body. **(1)**

(c) The painkiller is known to have serious side effects if there is more than 450 milligrams in the body.

Will the patient suffer side effects if this dose of painkiller is administered over a long time? **(4)**

4. (a) Express $2\cos x° - \sin x°$ in the form $k\sin(x+\alpha)°$, where $k > 0$ and $0 \le \alpha \le 360$. **(4)**

(b) Hence find algebraically the values of x between 0 and 360 for which $2\cos x° - \sin x° = 1$. **(4)**

5. The diagram which follows illustrates the graph of $y = f(x)$ where $f(x) = -x^3 + 3x^2 + x - 3$.

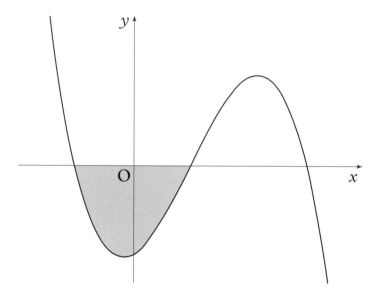

(a) Show that $(x-1)$ is a factor of $f(x)$. **(2)**

(b) Hence factorise $f(x)$ fully and determine the values of x such that $f(x) = 0$. **(3)**

(c) Calculate the shaded area. **(5)**

6. As shown in the following diagram, a set of experimental results gives rise to a straight line graph when $\log_{10} y$ is plotted against $\log_{10} x$.

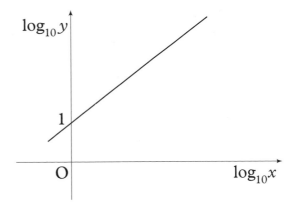

This straight line passes through the point $(0, 1)$ and has a gradient of 3.

Express y in terms of x. **(6)**

7. Find the values of p such that the equation $(p-1)x^2 + 4x + (2p-4) = 0$ has equal roots. **(4)**

8. A jewellery box, in the shape of a cuboid, is constructed with two square ends, as shown in the diagram.

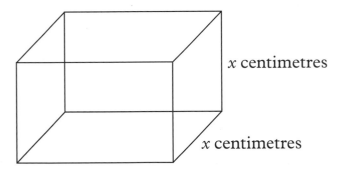

The total surface area of the cuboid is 216 square centimetres and the length of each square end is x centimetres.

(a) If the volume of the jewellery box is given as V cubic centimetres, show that

$$V = x\left(54 - \frac{1}{2}x^2\right).$$ **(3)**

(b) Find the dimensions of the jewellery box with maximum volume.

Justify your answer. **(6)**

[END OF QUESTION PAPER]

Answer Support for Paper Two A–D

Practice Paper A

PAPER TWO

1. (a) $\overrightarrow{AB} = \mathbf{b} - \mathbf{a} = \begin{pmatrix} 3 \\ 0 \\ -2 \end{pmatrix} - \begin{pmatrix} 1 \\ -6 \\ 2 \end{pmatrix} = \begin{pmatrix} 2 \\ 6 \\ -4 \end{pmatrix}$. $\overrightarrow{BC} = \mathbf{c} - \mathbf{b} = \begin{pmatrix} 6 \\ 9 \\ -8 \end{pmatrix} - \begin{pmatrix} 3 \\ 0 \\ -2 \end{pmatrix} = \begin{pmatrix} 3 \\ 9 \\ -6 \end{pmatrix}$.

Hence $\overrightarrow{AB} = \dfrac{2}{3} \overrightarrow{BC}$.

A, B and C are collinear because $\overrightarrow{AB}$ and $\overrightarrow{BC}$ have the same direction and a point in common.

(b) B divides AC in the ratio 2 : 3.

4 marks

2. (a) (i) $h(x) = f(g(x)) = f(x^2 + 2) = 2(x^2 + 2) - 3 = 2x^2 + 4 - 3 = 2x^2 + 1$.

(ii) $k(x) = g(f(x)) = g(2x - 3) = (2x - 3)^2 + 2 = 4x^2 - 12x + 9 + 2$

$= 4x^2 - 12x + 11$.

(b) $k(a) = 2h(a) - 15 \Rightarrow 4a^2 - 12a + 11 = 2(2a^2 + 1) - 15$.

Hence $4a^2 - 12a + 11 = 4a^2 + 2 - 15 \Rightarrow -12a = -13 - 11 = -24 \Rightarrow a = 2$.

5 marks

3. Use List of Formulae: $\cos 2A = 2\cos^2 A - 1$

$$2\cos 2x^\circ - 4\cos x^\circ - 1 = 0$$

$$\Rightarrow 2(2\cos^2 x^\circ - 1) - 4\cos x^\circ - 1 = 0$$

$$\Rightarrow 4\cos^2 x^\circ - 2 - 4\cos x^\circ - 1 = 0$$

$$\Rightarrow 4\cos^2 x^\circ - 4\cos x^\circ - 3 = 0$$

$$\Rightarrow (2\cos x^\circ + 1)(2\cos x^\circ - 3) = 0$$

$$\Rightarrow \cos x^\circ = -\frac{1}{2} \quad \text{or} \quad \cos x^\circ = \frac{3}{2}\,(!)$$

$$\Rightarrow x = 120 \text{ or } 240 \,\left(\cos x^\circ = \frac{3}{2} \text{ is an \textbf{impossible} solution}\right).$$

6 marks

4. (a) A function is decreasing when $f'(x) < 0$.

$f(x) = x^3 + 9x^2 + 27x \Rightarrow f'(x) = 3x^2 + 18x + 27 = 3(x^2 + 6x + 9) = 3(x+3)^2.$

For all values of x, $(x+3)^2 \geq 0$.

Therefore $f'(x) = 3(x+3)^2$ is always greater than or equal to zero.

Hence $f(x)$ is never decreasing.

(b) $\displaystyle\int_1^8 \left(\frac{x-2}{\sqrt[3]{x}}\right)dx = \int_1^8 \left(\frac{x-2}{x^{\frac{1}{3}}}\right)dx = \int_1^8 \left(\frac{x}{x^{\frac{1}{3}}} - \frac{2}{x^{\frac{1}{3}}}\right)dx = \int_1^8 \left(x^{\frac{2}{3}} - 2x^{-\frac{1}{3}}\right)dx$

$\displaystyle\int_1^8 \left(x^{\frac{2}{3}} - 2x^{-\frac{1}{3}}\right)dx = \left[\left(\frac{3}{5}x^{\frac{5}{3}} - 3x^{\frac{2}{3}}\right)\right]_1^8 = \left[\left(\frac{3}{5} \times \sqrt[3]{8^5} - 3 \times \sqrt[3]{8^2}\right) - \left(\frac{3}{5} - 3\right)\right]$

$$= \left(\frac{96}{5} - 12 - \frac{3}{5} + 3\right) = \frac{48}{5}.$$

10 marks

5.

$$\frac{1}{3}\begin{array}{|cccc} 6 & -17 & k & 12 \\[2mm] & 2 & -5 & \frac{1}{3}(k-5) \\[1mm] \hline 6 & -15 & (k-5) & \vdots\ \ 0 \end{array}$$

Hence $12+\dfrac{1}{3}(k-5)=0 \Rightarrow \dfrac{1}{3}(k-5)=-12 \Rightarrow k-5=-36 \Rightarrow k=-31$.

Hence $6x^3-17x^2-31x+12=\left(x-\dfrac{1}{3}\right)(6x^2-15x-36)=\left(x-\dfrac{1}{3}\right)\times 3(2x^2-5x-12)$

$\Rightarrow 6x^3-17x^2-31x+12=(3x-1)(2x^2-5x-12)$

$\qquad\qquad\qquad\qquad\qquad =(3x-1)(2x+3)(x-4).$

5 marks

---

6. (a) Use List of Formulae: $\mathbf{a}.\mathbf{b}=|\mathbf{a}||\mathbf{b}|\cos\theta$, where θ is the angle between $\mathbf{a}$ and $\mathbf{b}$.

$\mathbf{a}.(\mathbf{b}+\mathbf{c})=\mathbf{a}.\mathbf{b}+\mathbf{a}.\mathbf{c}=|\mathbf{a}||\mathbf{b}|\cos 60° + |\mathbf{a}||\mathbf{c}|\cos 90° = 6\times 5\times\dfrac{1}{2}+0=15.$

(b)

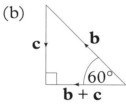

In this triangle,

$$\cos 60^{\mathrm{o}} = \frac{|\mathbf{b}+\mathbf{c}|}{|\mathbf{b}|} = \frac{|\mathbf{b}+\mathbf{c}|}{5} \Rightarrow |\mathbf{b}+\mathbf{c}|=5\times\cos 60^{\mathrm{o}} = 5\times 0\cdot 5 = 2\cdot 5.$$

5 marks

7. (a) Use List of Formulae: $\cos(A+B)=\cos A\cos B-\sin A\sin B$.

$$8\cos x^\circ - 15\sin x^\circ = k\cos(x+\alpha)^\circ$$
$$= k\cos x^\circ \cos\alpha^\circ - k\sin x^\circ \sin\alpha^\circ.$$

Hence $k\cos\alpha^\circ = 8$ and $k\sin\alpha^\circ = 15$.

$$k=\sqrt{8^2+15^2}=17.$$

$$\tan\alpha^\circ = \frac{15}{8} \Rightarrow \alpha = 61\cdot9.$$

NOTE: α is in the 1st quadrant because sine and cosine are both positive.

Hence $8\cos x^\circ - 15\sin x^\circ = 17\cos(x+61\cdot9)^\circ$.

(b) $8\cos x^\circ - 15\sin x^\circ = 12$

$$\Rightarrow 17\cos(x+61\cdot9)^\circ = 12$$

$$\Rightarrow \cos(x+61\cdot9)^\circ = \frac{12}{17}$$

$$\Rightarrow x+61\cdot9 = 45\cdot1 \text{ or } 360-45\cdot1$$

$$\Rightarrow x = -16\cdot8 \text{ or } 253 = -16\cdot8 + 360 \text{ or } 253 = 253 \text{ or } 343\cdot2.$$

(c) We solve this inequality using the results in part (b) along with a sketch of the graphs of $f(x)=17\cos(x+61\cdot9)^\circ$ and $f(x)=12$.

(The graph of $f(x)=17\cos(x+61\cdot9)^\circ$ is drawn by shifting the graph of $f(x)=17\cos x^\circ$ by $61\cdot9^\circ$ to the left.) Thus:

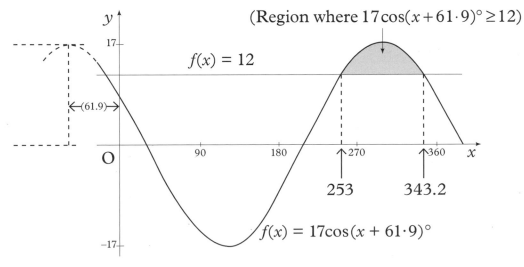

Hence solution is $253 \le x \le 343\cdot2$.

10 marks

8. (a) $T_t = T_0 e^{-kt} \Rightarrow 80 = 100 \times e^{-2k}$ (when $T_0 = 100$, $T_t = 80$ and $t = 2$).

Now take logarithms to the base e of both sides.

Hence $\ln 80 = \ln(100 \times e^{-2k}) = \ln 100 + \ln e^{-2k} = \ln 100 - 2k \ln e$.

As $\ln e = 1$, $\ln 80 = \ln 100 - 2k \Rightarrow k = \dfrac{\ln 100 - \ln 80}{2} = 0 \cdot 11157$.

(b) You can take the original temperature T_0 as 100, so that $T_t = 50$.

Hence $T_t = T_0 e^{-kt} \Rightarrow 50 = 100 \times e^{-0 \cdot 11157t}$.

Again take logarithms to the base e of both sides.

Therefore $\ln 50 = \ln(100 \times e^{-0 \cdot 11157t}) = \ln 100 + \ln e^{-0 \cdot 11157t} = \ln 100 - 0 \cdot 11157t \ln e$.

As $\ln e = 1$, $\ln 50 = \ln 100 - 0 \cdot 11157t \Rightarrow t = \dfrac{\ln 100 - \ln 50}{0 \cdot 11157} = 6 \cdot 2$.

So it takes $6 \cdot 2$ minutes for the liquid to cool to half its original temperature.

NOTE: In part (b), which is known as a 'half-life' question, it is possible to pick your own initial value and related half-life value. If you do so, a good choice is $T_0 = 100$ with $T_t = 50$.

6 marks

9. Use List of Formulae:

The equation $x^2 + y^2 + 2gx + 2fy + c = 0$ represents a circle with centre $(-g, -f)$ and radius $\sqrt{g^2 + f^2 - c}$.

(a) Centre C = $(-7, -2)$.

Radius = $\sqrt{g^2 + f^2 - c} = \sqrt{7^2 + 2^2 - 19} = \sqrt{34}$.

(b) Gradient of CA = $\dfrac{y_2 - y_1}{x_2 - x_1} = \dfrac{1 + 2}{-12 + 7} = \dfrac{3}{-5} = -\dfrac{3}{5}$.

Hence gradient of perpendicular line (the tangent at A) = $\dfrac{5}{3}$.

Equation of tangent is $y - b = m(x - a) \Rightarrow y - 1 = \dfrac{5}{3}(x + 12) \Rightarrow 3y - 3 = 5x + 60$.

This simplifies to $3y = 5x + 63$.

(c) To show that B (−9, 6) lies on the tangent, substitute $x = -9$ into equation.

$3y = 5 \times (-9) + 63 = -45 + 63 = 18 \Rightarrow y = 6$.

Hence B (−9, 6) lies on the tangent.

(d) Use List of Formulae:

The equation $(x - a)^2 + (y - b)^2 = r^2$ represents a circle with centre (a, b) and radius r.

Angle CAB = 90° (angle between tangent and radius), hence CB is a diameter of the circle.

Centre is midpoint of CB = $\left(\dfrac{-7 + (-9)}{2}, \dfrac{-2 + 6}{2} \right) = (-8, 2)$.

Use distance formula to find diameter:

$$d = \sqrt{(x_2 - x_1)^2 + (y_2 - y_1)^2} = \sqrt{(-9 + 7)^2 + (6 + 2)^2} = \sqrt{(-2)^2 + 8^2} = \sqrt{68}.$$

Hence radius = $\dfrac{1}{2}\sqrt{68}$.

Hence equation of circle is $(x + 8)^2 + (y - 2)^2 = \left(\dfrac{1}{2}\sqrt{68} \right)^2$.

This simplifies to $(x + 8)^2 + (y - 2)^2 = 17$.

9 marks

Answer Support for Paper Two A–D

Practice Paper B

PAPER TWO

1. (a) Use the section formula:

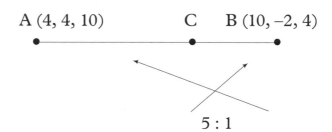

A (4, 4, 10)   C   B (10, −2, 4)

5 : 1

C is $\left(\dfrac{1\times4+5\times10}{6}, \dfrac{1\times4+5\times(-2)}{6}, \dfrac{1\times10+5\times4}{6}\right) = (9, -1, 5).$

(Alternatively x_A to x_B is 4 to 10 (=6). Point C lies $\dfrac{5}{6}$ of the way, so

$x_C = 4 + \dfrac{5}{6}(6) = 4 + 5 = 9.$)

(b) $\overrightarrow{CD} = \mathbf{d} - \mathbf{c} = \begin{pmatrix} x \\ -2 \\ 2 \end{pmatrix} - \begin{pmatrix} 9 \\ -1 \\ 5 \end{pmatrix} = \begin{pmatrix} x-9 \\ -1 \\ -3 \end{pmatrix}$, $\overrightarrow{AB} = \mathbf{b} - \mathbf{a} = \begin{pmatrix} 10 \\ -2 \\ 4 \end{pmatrix} - \begin{pmatrix} 4 \\ 4 \\ 10 \end{pmatrix} = \begin{pmatrix} 6 \\ -6 \\ -6 \end{pmatrix}.$

A (4, 4, 10)   C (9, −1, 5)   B (10, −2, 4)

D (x, −2, 2)

Use List of Formulae:

$\mathbf{a}.\mathbf{b} = a_1b_1 + a_2b_2 + a_3b_3$ where $\mathbf{a} = \begin{pmatrix} a_1 \\ a_2 \\ a_3 \end{pmatrix}$ and $\mathbf{b} = \begin{pmatrix} b_1 \\ b_2 \\ b_3 \end{pmatrix}.$

If CD is perpendicular to AB, then $\overrightarrow{CD}.\overrightarrow{AB} = 0.$

Hence $6(x-9) + (-1)\times(-6) + (-3)\times(-6) = 0 \Rightarrow 6x - 54 + 6 + 18 = 0.$

Therefore $6x = 30$ and $x = 5.$

7 marks

2. (a) Use List of Formulae:

The equation $x^2 + y^2 + 2gx + 2fy + c = 0$ represents a circle with centre $(-g, -f)$.

Hence C is the point $(3, 5)$.

(b) The chord with midpoint $(4, 6)$ is perpendicular to CM.

The gradient of CM is $\dfrac{y_2 - y_1}{x_2 - x_1} = \dfrac{6-5}{4-3} = \dfrac{1}{1} = 1$.

Therefore the gradient of the chord is -1.

The equation of the chord is $y - b = m(x - a) \Rightarrow y - 6 = -1(x - 4) \Rightarrow y = -x + 10$.

(c) We solve a system of equations to show that $x + y = 12$ is a tangent to the circle $x^2 + y^2 - 6x - 10y + 26 = 0$.

$$x + y = 12 \Rightarrow x = 12 - y \Rightarrow (12 - y)^2 + y^2 - 6(12 - y) - 10y + 26 = 0$$
$$\Rightarrow 144 - 24y + y^2 + y^2 - 72 + 6y - 10y + 26 = 0$$
$$\Rightarrow 2y^2 - 28y + 98 = 0$$
$$\Rightarrow y^2 - 14y + 49 = 0$$
$$\Rightarrow (y - 7)(y - 7) = 0.$$

NOTE: In the equation $y^2 - 14y + 49 = 0$, $b^2 - 4ac = (-14)^2 - 4 \times 1 \times 49 = 0$.

This indicates equal roots, hence the line is a tangent to the circle.

By solving the equation, $y = 7$, hence $x = 5 \Rightarrow$ P is $(5, 7)$.

10 marks

3. Use List of Formulae: $\sin 2A = 2\sin A\cos A$.

$$4\sin 2x° - 6\cos x° = 0 \Rightarrow 4(2\sin x° \cos x°) - 6\cos x° = 0$$

$$\Rightarrow 8\sin x° \cos x° - 6\cos x° = 0$$

$$\Rightarrow 2\cos x°(4\sin x° - 3) = 0$$

$$\Rightarrow \cos x° = 0 \text{ or } \sin x° = \frac{3}{4}$$

$$\Rightarrow x = 90, 270, 48·6, 131·4$$

$$\Rightarrow x = 48·6, 90, 131·4, 270.$$

5 marks

4. (a) Use List of Formulae: $\cos(A - B) = \cos A\cos B + \sin A\sin B$.

$$\cos\theta - \sqrt{3}\sin\theta = k\cos(\theta - \alpha)$$

$$= k\cos\theta\cos\alpha + k\sin\theta\sin\alpha.$$

Hence $k\cos\alpha = 1$ and $k\sin\alpha = -\sqrt{3}$.

$$k = \sqrt{1^2 + \left(-\sqrt{3}\right)^2} = 2.$$

$$\tan\alpha = -\frac{\sqrt{3}}{1} = -\sqrt{3} \Rightarrow \alpha = \frac{5\pi}{3}.$$

NOTE: α is in the 4th quadrant because sine is negative and cosine is positive.

Hence $\cos\theta - \sqrt{3}\sin\theta = 2\cos\left(\theta - \frac{5\pi}{3}\right)$.

(b) Use List of Formulae: If $f(x) = \cos ax$, then $f'(x) = -a\sin ax$.

Let $f(\theta) = \cos\theta - \sqrt{3}\sin\theta = 2\cos\left(\theta - \frac{5\pi}{3}\right)$.

Hence the derivative, $f'(\theta) = -2\sin\left(\theta - \frac{5\pi}{3}\right)$.

6 marks

5. (a) This is a very difficult proof for most students. Remember that if you cannot complete it, use the correct answer to do part (b).

Area, A = length × breadth. The length of the garden is $(x + 4)$ metres.

What is the breadth? Use the fact that the area of the lawn is 72 square metres.

Then as area of lawn = 72, breadth of lawn = $\dfrac{72}{x}$.

Hence breadth of garden = $\dfrac{72}{x} + 2$. We already know that the length of the garden is $(x + 4)$.

Hence $A = (x+4)\left(\dfrac{72}{x}+2\right) = 72 + 2x + \dfrac{288}{x} + 8 = 80 + 2x + \dfrac{288}{x}$.

(b) This is an optimisation problem. Equate the derivative of A to zero and proceed.

$A = 80 + 2x + 288x^{-1} \Rightarrow \dfrac{dA}{dx} = 2 - 288x^{-2} = 2 - \dfrac{288}{x^2} = 0$ at stationary points.

$2 - \dfrac{288}{x^2} = 0 \Rightarrow 2x^2 - 288 = 0 \Rightarrow x^2 = \dfrac{288}{2} = 144 \Rightarrow x = \pm\sqrt{144} = \pm 12.$

Hence stationary points occur when $x = \pm 12$.

NOTE: A negative answer (−12) is not possible.

Nature table:

x	$\rightarrow$	−12	$\rightarrow$	12	$\rightarrow$
$\dfrac{dA}{dx}$	+	0	−	0	+
	/	max	\	min	/

Hence the stationary point is a minimum turning point when $x = 12$.

Therefore the dimensions of the garden with minimum area are 16 metres (length) by 8 metres (breadth).

9 marks

6. (a) (i) $f(h(x)) = f\left(x - \dfrac{\pi}{3}\right) = \sin\left(x - \dfrac{\pi}{3}\right).$

(ii) $g(k(x)) = g\left(x - \dfrac{\pi}{6}\right) = \cos\left(x - \dfrac{\pi}{6}\right).$

(b) (i) Use List of Formulae: $\sin(A - B) = \sin A \cos B - \cos A \sin B.$

$$f(h(x)) = \sin\left(x - \frac{\pi}{3}\right) = \sin x \cos \frac{\pi}{3} - \cos x \sin \frac{\pi}{3}$$

$$= \frac{1}{2}\sin x - \frac{\sqrt{3}}{2}\cos x.$$

(ii) Use List of Formulae: $\cos(A - B) = \cos A \cos B + \sin A \sin B.$

$$g(k(x)) = \cos\left(x - \frac{\pi}{6}\right) = \cos x \cos \frac{\pi}{6} + \sin x \sin \frac{\pi}{6}$$

$$= \frac{\sqrt{3}}{2}\cos x + \frac{1}{2}\sin x.$$

Hence $f(h(x)) - g(k(x)) = \dfrac{1}{2}\sin x - \dfrac{\sqrt{3}}{2}\cos x - \left(\dfrac{\sqrt{3}}{2}\cos x + \dfrac{1}{2}\sin x\right)$

$$= -\sqrt{3}\cos x.$$

Hence $f(h(x)) - g(k(x)) = \dfrac{\sqrt{3}}{2} \Rightarrow -\sqrt{3}\cos x = \dfrac{\sqrt{3}}{2} \Rightarrow \cos x = -\dfrac{1}{2}.$

Therefore $x = \dfrac{2\pi}{3}$ or $x = \dfrac{4\pi}{3}.$

9 marks

7. (a) $\log_{12} x + \log_{12} y = 1 \Rightarrow \log_{12} xy = 1 \Rightarrow 12^1 = xy \Rightarrow xy = 12 \Rightarrow x = \dfrac{12}{y}.$

    (b) Substitute $x = \dfrac{12}{y}$ into $4x + y = 19 \Rightarrow 4 \times \dfrac{12}{y} + y = 19.$

    Therefore $\dfrac{48}{y} + y = 19 \Rightarrow 48 + y^2 = 19y \Rightarrow y^2 - 19y + 48 = 0.$

    Therefore $(y - 3)(y - 16) = 0 \Rightarrow y = 3$ or $16.$

    When $y = 3$, $x = 4$; when $y = 16$, $x = \dfrac{12}{16} = \dfrac{3}{4}.$

    Solutions are $(4, 3)$ or $\left(\dfrac{3}{4}, 16\right).$

6 marks

8. To calculate the shaded area, integration must be used.

    Several strategies may be employed, but it would be simplest to calculate the area to the right of the y-axis and double it as the y-axis is an axis of symmetry for the whole shaded area.

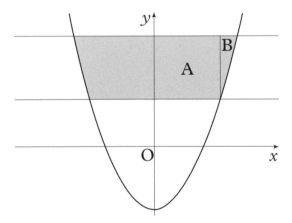

    Start by calculating limits.

    $y = x^2 - 9$ meets $y = 16$ where $x^2 - 9 = 16 \Rightarrow x^2 = 25 \Rightarrow x = \pm 5.$

    $y = x^2 - 9$ meets $y = 7$ where $x^2 - 9 = 7 \Rightarrow x^2 = 16 \Rightarrow x = \pm 4.$

    Hence rectangular shaded area, A $= (16 - 7) \times 4 = 36.$

Area B, between $y = 16$ and $y = x^2 - 9$ is $\displaystyle\int_4^5 \left[16 - (x^2 - 9)\right] dx = \int_4^5 \left(25 - x^2\right) dx$.

Hence area $= \left[25x - \dfrac{1}{3}x^3\right]_4^5 = \left[25 \times 5 - \dfrac{1}{3} \times 5^3\right] - \left[25 \times 4 - \dfrac{1}{3} \times 4^3\right]$

$= \left(125 - \dfrac{125}{3} - 100 + \dfrac{64}{3}\right) = \dfrac{14}{3}$.

Hence total shaded area $= 2 \times (A + B) = 2 \times \left(36 + \dfrac{14}{3}\right) = 81\tfrac{1}{3}$.

8 marks

Answer Support for Paper Two A–D

Practice Paper C

PAPER TWO

---

1. (a) Solve a system of equations to show that $y - 5x + 24 = 0$ is a tangent to the parabola $y = x^2 - 3x - 8$.

$$y - 5x + 24 = 0 \Rightarrow y = 5x - 24 \Rightarrow 5x - 24 = x^2 - 3x - 8$$

$$\Rightarrow x^2 - 3x - 5x - 8 + 24 = 0$$

$$\Rightarrow x^2 - 8x + 16 = 0$$

$$\Rightarrow (x - 4)(x - 4) = 0.$$

NOTE: In the equation $x^2 - 8x + 16 = 0$, $b^2 - 4ac = (-8)^2 - 4 \times 1 \times 16 = 0$.

This indicates equal roots, hence the line is a tangent to the parabola.

(b) By solving the equation $(x - 4)(x - 4) = 0$, $x = 4$, hence $y \ (= 5x - 24) = -4$ $\Rightarrow$ point of contact is $(4, -4)$.

5 marks

---

2. (a) $\overrightarrow{DF} = \mathbf{f} - \mathbf{d} = \begin{pmatrix} 6 \\ -3 \\ 5 \end{pmatrix} - \begin{pmatrix} 2 \\ -1 \\ 1 \end{pmatrix} = \begin{pmatrix} 4 \\ -2 \\ 4 \end{pmatrix}$. $\overrightarrow{FE} = \mathbf{e} - \mathbf{f} = \begin{pmatrix} 8 \\ -4 \\ 7 \end{pmatrix} - \begin{pmatrix} 6 \\ -3 \\ 5 \end{pmatrix} = \begin{pmatrix} 2 \\ -1 \\ 2 \end{pmatrix}$.

Hence $\overrightarrow{DF} = 2\overrightarrow{FE}$.

D, F and E are collinear as $\overrightarrow{DF}$ and $\overrightarrow{FE}$ have the same direction and F is a common point.

F divides DE in the ratio 2 : 1.

(b) Since DP is perpendicular to EP, then $\overrightarrow{DP}.\overrightarrow{EP} = 0$.

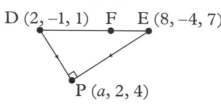

D (2, –1, 1)   F   E (8, –4, 7)

P (a, 2, 4)

$$\overrightarrow{DP} = \mathbf{p} - \mathbf{d} = \begin{pmatrix} a \\ 2 \\ 4 \end{pmatrix} - \begin{pmatrix} 2 \\ -1 \\ 1 \end{pmatrix} = \begin{pmatrix} a-2 \\ 3 \\ 3 \end{pmatrix}. \quad \overrightarrow{EP} = \mathbf{p} - \mathbf{e} = \begin{pmatrix} a \\ 2 \\ 4 \end{pmatrix} - \begin{pmatrix} 8 \\ -4 \\ 7 \end{pmatrix} = \begin{pmatrix} a-8 \\ 6 \\ -3 \end{pmatrix}.$$

Use List of Formulae: $\mathbf{a}.\mathbf{b} = a_1b_1 + a_2b_2 + a_3b_3$ where $\mathbf{a} = \begin{pmatrix} a_1 \\ a_2 \\ a_3 \end{pmatrix}$ and $\mathbf{b} = \begin{pmatrix} b_1 \\ b_2 \\ b_3 \end{pmatrix}$.

Hence $\overrightarrow{DP}.\overrightarrow{EP} = (a-2)(a-8) + 3 \times 6 + 3 \times (-3) = (a-2)(a-8) + 18 - 9 = 0$.

Therefore $a^2 - 8a - 2a + 16 + 18 - 9 = 0 \Rightarrow a^2 - 10a + 25 = 0$.

This leads to $(a-5)^2 = 0$, so $a = 5$.

(c) Use the distance formula: $d = \sqrt{(x_2 - x_1)^2 + (y_2 - y_1)^2 + (z_2 - z_1)^2}$.

DF $= \sqrt{(6-2)^2 + (-3+1)^2 + (5-1)^2} = \sqrt{16+4+16} = \sqrt{36} = 6$.

DP $= \sqrt{(5-2)^2 + (2+1)^2 + (4-1)^2} = \sqrt{9+9+9} = \sqrt{27}$. (P is the point (5, 2, 4).)

FP $= \sqrt{(5-6)^2 + (2+3)^2 + (4-5)^2} = \sqrt{1+25+1} = \sqrt{27}$.

Since DP = FP, triangle DPF is isosceles.

Use List of Formulae: $\mathbf{a}.\mathbf{b} = |\mathbf{a}||\mathbf{b}|\cos\theta$, where θ is the angle between $\mathbf{a}$ and $\mathbf{b}$

or $\mathbf{a}.\mathbf{b} = a_1b_1 + a_2b_2 + a_3b_3$ where $\mathbf{a} = \begin{pmatrix} a_1 \\ a_2 \\ a_3 \end{pmatrix}$ and $\mathbf{b} = \begin{pmatrix} b_1 \\ b_2 \\ b_3 \end{pmatrix}$.

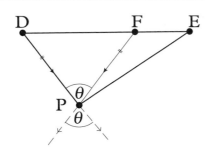

$$\overrightarrow{DP} = \begin{pmatrix} 3 \\ 3 \\ 3 \end{pmatrix}. \quad \overrightarrow{FP} = \mathbf{p} - \mathbf{f} = \begin{pmatrix} 5 \\ 2 \\ 4 \end{pmatrix} - \begin{pmatrix} 6 \\ -3 \\ 5 \end{pmatrix} = \begin{pmatrix} -1 \\ 5 \\ -1 \end{pmatrix}.$$

Here $\mathbf{a}.\mathbf{b} = \overrightarrow{DP}.\overrightarrow{FP} = (3 \times -1) + (3 \times 5) + (3 \times -1) = -3 + 15 + (-3) = 9.$

$$\cos\theta = \frac{\mathbf{a}.\mathbf{b}}{|\mathbf{a}||\mathbf{b}|} = \frac{9}{\sqrt{27} \times \sqrt{27}} = \frac{9}{27} = \frac{1}{3} \Rightarrow \theta = \cos^{-1}\left(\frac{1}{3}\right) = 70 \cdot 5^{\circ}.$$

Therefore angle DPF is $70 \cdot 5^{\circ}$.

14 marks

---

3. Use List of Formulae: $\cos 2A = 1 - 2\sin^2 A.$

$$\cos 2x^{\circ} - 4\sin x^{\circ} + 5 = 0$$

$$\Rightarrow 1 - 2\sin^2 x^{\circ} - 4\sin x^{\circ} + 5 = 0$$

$$\Rightarrow 2\sin^2 x^{\circ} + 4\sin x^{\circ} - 6 = 0$$

$$\Rightarrow \sin^2 x^{\circ} + 2\sin x^{\circ} - 3 = 0$$

$$\Rightarrow (\sin x^{\circ} + 3)(\sin x^{\circ} - 1) = 0$$

$$\Rightarrow \sin x^{\circ} = -3 \text{ or } \sin x^{\circ} = 1$$

$$\Rightarrow x = 90 \, (\sin x^{\circ} = -3 \text{ is an } \textbf{impossible} \text{ solution}).$$

6 marks

4. $y = \sqrt{x} + \dfrac{6}{\sqrt{x}} = x^{\frac{1}{2}} + \dfrac{6}{x^{\frac{1}{2}}} = x^{\frac{1}{2}} + 6x^{-\frac{1}{2}} \Rightarrow \dfrac{dy}{dx} = \dfrac{1}{2}x^{-\frac{1}{2}} - 3x^{-\frac{3}{2}} = \dfrac{1}{2x^{\frac{1}{2}}} - \dfrac{3}{x^{\frac{3}{2}}}.$

Hence gradient $= \dfrac{1}{2x^{\frac{1}{2}}} - \dfrac{3}{x^{\frac{3}{2}}} = \dfrac{1}{2\sqrt{x}} - \dfrac{3}{\sqrt{x^3}}.$

When $x = 4$, gradient $= \dfrac{1}{2\sqrt{x}} - \dfrac{3}{\sqrt{x^3}} = \dfrac{1}{2\sqrt{4}} - \dfrac{3}{\sqrt{4^3}} = \dfrac{1}{4} - \dfrac{3}{8} = -\dfrac{1}{8}.$

$8x - y = 5 \Rightarrow y = 8x - 5.$ This line has gradient $= 8$.

The lines are perpendicular since $8 \times \left(-\dfrac{1}{8}\right) = -1.$

5 marks

5. (a) Use List of Formulae: $\cos(A+B) = \cos A \cos B - \sin A \sin B.$

$5\cos x^\circ - 12\sin x^\circ = k\cos(x+\alpha)^\circ$

$\qquad\qquad\qquad\quad = k\cos x^\circ \cos\alpha^\circ - k\sin x^\circ \sin\alpha^\circ.$

Hence $k\cos\alpha^\circ = 5$ and $k\sin\alpha^\circ = 12.$

$k = \sqrt{5^2 + 12^2} = 13.$

$\tan\alpha^\circ = \dfrac{12}{5} \Rightarrow \alpha = 67\cdot4.$

NOTE: α is in the 1st quadrant because sine and cosine are both positive.

Hence $5\cos x^\circ - 12\sin x^\circ = 13\cos(x+67\cdot4)^\circ.$

(b) The maximum value of $5\cos x^\circ - 12\sin x^\circ$ is 13.

This occurs when $\cos(x+67\cdot4)^\circ = 1$, i.e. when $x + 67\cdot4 = 360$ and hence

$x = 360 - 67\cdot4 = 292\cdot6.$

The minimum value of $5\cos x^\circ - 12\sin x^\circ$ is -13.

This occurs when $\cos(x+67\cdot4)^\circ = -1$, i.e. when $x + 67\cdot4 = 180$ and hence

$x = 180 - 67\cdot4 = 112\cdot6.$

(c) $y = 5\cos x° - 12\sin x°$ crosses the y-axis when $x = 0$.

This occurs when $y = 5\cos 0° - 12\sin 0° = 5$, i.e. at $(0, 5)$.

To sketch the graph of $y = 5\cos x° - 12\sin x°$, sketch $y = 13\cos(x + 67·4)°$.

Then think of the graph of $y = 13\cos x°$ moved $67·4°$ to the left.

(This is the same as moving it $(360 - 67·4)°$ to the right. That is, $292·6°$ to the right.)

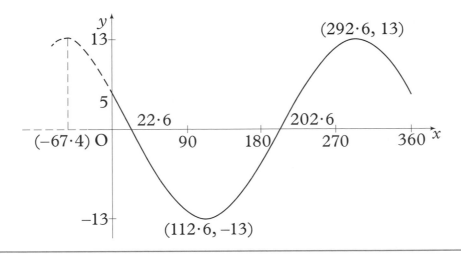

10 marks

6. Replace x with 1 in the equation $y = x^3 - x + 3 \Rightarrow y = 1^3 - 1 + 3 = 3$.

Hence the point $(1, 3)$ lies on the curve $y = x^3 - x + 3$.

The line OP passes through the origin and the point $(1, 3)$, so its equation is $y = 3x$.

The shaded area $= \displaystyle\int_0^1 \left[(x^3 - x + 3 - 3x) \right] dx = \int_0^1 (x^3 - 4x + 3) dx$

$$= \left[\frac{1}{4}x^4 - 2x^2 + 3x \right]_0^1 = \left(\frac{1}{4} - 2 + 3 \right) = \frac{5}{4}.$$

6 marks

7. (a) Find the equation of the line CD.

The gradient of CD is $\dfrac{y_2 - y_1}{x_2 - x_1} = \dfrac{0-6}{18-0} = \dfrac{-6}{18} = -\dfrac{1}{3}$.

As the y-intercept is 6, the equation of CD is $y = -\dfrac{1}{3}x + 6$.

As E (q, p) lies on CD, $p = -\dfrac{1}{3}q + 6 \Rightarrow p = \dfrac{1}{3}(18 - q)$.

Area of rectangle, $A(q) = $ length $\times$ breadth $= qp$.

Hence $A(q) = q \times \dfrac{1}{3}(18 - q) = 6q - \dfrac{1}{3}q^2$.

NOTE: Part (a) could also be solved using similar triangles.

(b) This is an optimisation problem. Equate the derivative of $A(q)$ to zero and proceed.

$A(q) = 6q - \dfrac{1}{3}q^2 \Rightarrow A'(q) = 6 - \dfrac{2}{3}q = 0$ at stationary points.

$6 - \dfrac{2}{3}q = 0 \Rightarrow 18 - 2q = 0 \Rightarrow q = 9$.

Nature table:

q	$\rightarrow$	9	$\rightarrow$
$A'(q)$	+	0	−
	/	‾	\
		max	

Hence the stationary point is a maximum turning point when $q = 9$.

So the largest area occurs when $q = 9$.

8 marks

8. (a) Let $\log_{16} x = a \Rightarrow x = 16^a$. Let $\log_4 x = b \Rightarrow x = 4^b$.

As $x = 16^a$ and $x = 4^b$, then $16^a = 4^b \Rightarrow \left(4^2\right)^a = 4^b \Rightarrow 4^{2a} = 4^b$.

Hence $2a = b \Rightarrow 2\log_{16} x = \log_4 x \Rightarrow \log_{16} x = \frac{1}{2}\log_4 x$.

(b) $\log_{16} x = (\log_4 x)^2 \Rightarrow \frac{1}{2}\log_4 x = \left(\log_4 x\right)^2$ using the result from part (a).

Now divide both sides by $\log_4 x$.

This leads to $\frac{1}{2} = \log_4 x \Rightarrow x = 4^{\frac{1}{2}} = \sqrt{4} = 2$, hence $x = 2$.

6 marks

Answer Support for Paper Two A–D

Practice Paper D

PAPER TWO

---

1. (a) M is the point $(5, 6, 8)$.

(b) N is the point $(0, 0, 4)$.

$$\overrightarrow{MF} = \mathbf{f} - \mathbf{m} = \begin{pmatrix} 10 \\ 6 \\ 8 \end{pmatrix} - \begin{pmatrix} 5 \\ 6 \\ 8 \end{pmatrix} = \begin{pmatrix} 5 \\ 0 \\ 0 \end{pmatrix}. \quad \overrightarrow{MN} = \mathbf{n} - \mathbf{m} = \begin{pmatrix} 0 \\ 0 \\ 4 \end{pmatrix} - \begin{pmatrix} 5 \\ 6 \\ 8 \end{pmatrix} = \begin{pmatrix} -5 \\ -6 \\ -4 \end{pmatrix}.$$

(c) Use List of Formulae: $\mathbf{a}.\mathbf{b} = |\mathbf{a}||\mathbf{b}|\cos\theta$, where θ is the angle between $\mathbf{a}$ and $\mathbf{b}$

or $\mathbf{a}.\mathbf{b} = a_1 b_1 + a_2 b_2 + a_3 b_3$ where $\mathbf{a} = \begin{pmatrix} a_1 \\ a_2 \\ a_3 \end{pmatrix}$ and $\mathbf{b} = \begin{pmatrix} b_1 \\ b_2 \\ b_3 \end{pmatrix}$.

$$\overrightarrow{MF}.\overrightarrow{MN} = a_1 b_1 + a_2 b_2 + a_3 b_3 = 5 \times (-5) + 0 \times (-6) + 0 \times (-4) = -25.$$

$$\cos\theta = \frac{\mathbf{a}.\mathbf{b}}{|\mathbf{a}||\mathbf{b}|} = \frac{-25}{\sqrt{5^2} \times \sqrt{(-5)^2 + (-6)^2 + (-4)^2}} = \frac{-25}{5 \times \sqrt{77}} = -0.5698.$$

Therefore angle FMN is $124.7°$.

8 marks

---

2. Use List of Formulae: $f(x) = \sin ax \Rightarrow f'(x) = a\cos ax$.

$$f(x) = \cos ax \Rightarrow f'(x) = -a\sin ax.$$

$$\sin 2A = 2\sin A \cos A.$$

$f(x) = \sin 2x + \cos^2 x = \sin 2x + (\cos x)^2 \Rightarrow f'(x) = 2\cos 2x + 2\cos x \times (-\sin x)$.

Therefore $f'(x) = 2\cos 2x - 2\sin x \cos x$.

This simplifies to $f'(x) = 2\cos 2x - \sin 2x$ using the List of Formulae.

When $x = \dfrac{\pi}{4}$, the gradient of the tangent is $2\cos 2x - \sin 2x = 2\cos\dfrac{\pi}{2} - \sin\dfrac{\pi}{2}$.

$$2\cos\frac{\pi}{2} - \sin\frac{\pi}{2} = 2\times 0 - 1 = -1.$$

To find the point of contact, replace x with $\dfrac{\pi}{4}$ in $f(x) = \sin 2x + \cos^2 x$.

This leads to $f\left(\dfrac{\pi}{4}\right) = \sin\dfrac{\pi}{2} + \cos^2\dfrac{\pi}{4} = 1 + \left(\dfrac{1}{\sqrt{2}}\right)^2 = 1 + \dfrac{1}{2} = \dfrac{3}{2}$.

The point of contact is $\left(\dfrac{\pi}{4}, \dfrac{3}{2}\right)$.

The equation of the tangent is $y - b = m(x - a) \Rightarrow y - \dfrac{3}{2} = -1\left(x - \dfrac{\pi}{4}\right)$.

7 marks

3. **(a)** A reduction of 20% is the same as retaining 80%, a reduction factor of $0\cdot 8$.

$$75 \times 0\cdot 8 = 60 \qquad 60 + 75 = 135$$
$$135 \times 0\cdot 8 = 108 \qquad 108 + 75 = 183$$
$$183 \times 0\cdot 8 = 146 \qquad 146 + 75 = 221$$
$$221 \times 0\cdot 8 = 177$$

Therefore the painkiller becomes effective after the 4th dose.

(b) $u_{n+1} = 0\cdot 8u_n + 75$.

(c) Since $-1 < 0\cdot 8 < 1$, this sequence has a limit.

$$L = \frac{b}{1-a} = \frac{75}{1 - 0\cdot 8} = \frac{75}{0\cdot 2} = 375.$$

As $L = 375 < 450$, there are no side effects.

8 marks

4. (a) Use List of Formulae: $\sin(A+B) = \sin A \cos B + \cos A \sin B$.

$$2\cos x° - \sin x° = k\sin(x+\alpha)°$$
$$= k\sin x° \cos\alpha° + k\cos x° \sin\alpha°.$$

Hence $k\cos\alpha° = -1$ and $k\sin\alpha° = 2$.

$$k = \sqrt{(-1)^2 + 2^2} = \sqrt{5}.$$

$$\tan\alpha° = \frac{2}{-1} = -2 \Rightarrow \alpha = 116\cdot6.$$

NOTE: α is in the 2nd quadrant because sine is positive and cosine is negative.

Hence $2\cos x° - \sin x° = \sqrt{5}\sin(x+116\cdot6)°$.

(b) $\quad 2\cos x° - \sin x° = 1$

$$\Rightarrow \sqrt{5}\sin(x+116\cdot6)° = 1$$

$$\Rightarrow \sin(x+116\cdot6)° = \frac{1}{\sqrt{5}}$$

$$\Rightarrow x+116\cdot6 = 26\cdot6 \text{ or } 180 - 26\cdot6$$

$$\Rightarrow x = 26\cdot6 - 116\cdot6 \text{ or } 180 - 26\cdot6 - 116\cdot6$$

$$\Rightarrow x = -90 \text{ or } 36\cdot8$$

$$\Rightarrow x = -90 + 360 \text{ or } 36\cdot8$$

$$\Rightarrow x = 36\cdot8 \text{ or } 270.$$

8 marks

5. (a)

$$
\begin{array}{r|rrrr}
1 & -1 & 3 & 1 & -3 \\
 & & -1 & 2 & 3 \\
\hline
 & -1 & 2 & 3 & 0 \\
\end{array}
$$

Hence $(x-1)$ is a factor of $f(x)$.

(b) $f(x) = -x^3 + 3x^2 + x - 3 = (x-1)(-x^2 + 2x + 3) = (x-1)(3-x)(1+x)$.

Hence $f(x) = 0$ when $x = -1, 1, 3$.

(c) Shaded area $= \displaystyle\int_{-1}^{1} \left(-x^3 + 3x^2 + x - 3\right) dx$

$$
= \left[-\frac{1}{4}x^4 + x^3 + \frac{1}{2}x^2 - 3x \right]_{-1}^{1}
$$

$$
= \left(-\frac{1}{4} + 1 + \frac{1}{2} - 3 \right) - \left(-\frac{1}{4} - 1 + \frac{1}{2} + 3 \right)
$$

$$
= -\frac{7}{4} - \frac{9}{4} = -\frac{16}{4} = -4.
$$

Result is negative as the area is below the x-axis.

So shaded area $= 4$.

10 marks

6. Let $X = \log_{10} x$ and $Y = \log_{10} y$.

As the straight line has gradient 3 and Y-intercept 1, its equation is $Y = 3X + 1$.

Hence $\log_{10} y = 3\log_{10} x + 1$

$\Rightarrow \qquad \log_{10} y = 3\log_{10} x + \log_{10} 10$

$\Rightarrow \qquad \log_{10} y = \log_{10} x^3 + \log_{10} 10$

$\Rightarrow \qquad \log_{10} y = \log_{10} 10x^3$

$\Rightarrow \qquad y = 10x^3.$

6 marks

7. The equation $(p-1)x^2 + 4x + (2p-4) = 0$ has equal roots when $b^2 - 4ac = 0$.

   Hence for equal roots, $4^2 - 4 \times (p-1) \times (2p-4) = 0$

   $$\Rightarrow 16 - 4(2p^2 - 6p + 4) = 0$$

   $$\Rightarrow 16 - 8p^2 + 24p - 16 = 0$$

   $$\Rightarrow -8p^2 + 24p = 0$$

   $$\Rightarrow 8p^2 - 24p = 0$$

   $$\Rightarrow 8p(p-3) = 0$$

   $$\Rightarrow p = 0 \text{ or } 3.$$

4 marks

8. (a) Let the length of the cuboid be l.

   First find a formula for the surface area, A, of the cuboid and then change the subject of the formula to l.

   $$A = 2x^2 + 4lx \Rightarrow 4lx = A - 2x^2 \Rightarrow l = \frac{A - 2x^2}{4x} = \frac{216 - 2x^2}{4x}, \text{ as } A = 216.$$

   Volume of cuboid, V = length $\times$ breadth $\times$ height

   $$= lx^2$$

   $$= \left(\frac{216 - 2x^2}{4x}\right) \times x^2$$

   $$= \frac{216x^2}{4x} - \frac{2x^4}{4x}$$

   $$= 54x - \frac{1}{2}x^3$$

   $$= x\left(54 - \frac{1}{2}x^2\right).$$

   HINT: In proofs of this type, use the information given, in this case the surface area. If you have to introduce a new variable, l, to create a formula, change the subject of the formula to l, and use this in completing the proof.

(b) This is an optimisation problem. Equate the derivative of V to zero and proceed.

$$V = x\left(54 - \frac{1}{2}x^2\right) = 54x - \frac{1}{2}x^3 \Rightarrow \frac{dV}{dx} = 54 - \frac{3}{2}x^2 = 0 \text{ at stationary points.}$$

Hence $54 - \frac{3}{2}x^2 = 0 \Rightarrow 108 - 3x^2 = 0 \Rightarrow 3x^2 = 108 \Rightarrow x^2 = 36 \Rightarrow x = \pm\sqrt{36} = \pm 6.$

Hence stationary points occur when $x = -6$ or 6.

Nature table:

x	$\rightarrow$	-6	$\rightarrow$	6	$\rightarrow$
$\dfrac{dV}{dx}$	$-$	0	$+$	0	$-$
	$\searrow$	min	$\nearrow$	max	$\searrow$

The result $x = -6$ is impossible.

Hence the stationary point is a maximum turning point when $x = 6$.

When $x = 6$, $l = \dfrac{A - 2x^2}{4x} = \dfrac{216 - 2 \times 6^2}{4 \times 6} = 6.$

Therefore the dimensions of the jewellery box with maximum volume are 6 centimetres by 6 centimetres by 6 centimetres.

9 marks

Appendix 1

Useful Facts and Formulae

The List of Formulae

The List of Formulae at the start of the exam paper is an invaluable aid if used properly. It is important that you refer to it regularly during your exam. Even if you feel that you know all the formulae involved, check the list to ensure accuracy. Also take care when copying formulae from the list. It is easy to make a transcription error when copying a formula so always double check.

It is essential, too, that you know when to refer to the list. Here are some recommendations:

1. If you need to find the centre or radius of a circle, use the formula $x^2 + y^2 + 2gx + 2fy + c = 0$.

2. If you are asked to find the equation of a circle, use the formula $(x-a)^2 + (y-b)^2 = r^2$.

   (In this case, you may have to use the distance formula to calculate the radius.)

3. In questions on vectors requiring you to calculate the size of an angle, use the formulae

   $$\mathbf{a}.\mathbf{b} = |\mathbf{a}||\mathbf{b}| \cos\theta, \text{ where } \theta \text{ is the angle between } \mathbf{a} \text{ and } \mathbf{b}$$

   or $\mathbf{a}.\mathbf{b} = a_1 b_1 + a_2 b_2 + a_3 b_3$ where $\mathbf{a} = \begin{pmatrix} a_1 \\ a_2 \\ a_3 \end{pmatrix}$ and $\mathbf{b} = \begin{pmatrix} b_1 \\ b_2 \\ b_3 \end{pmatrix}$.

4. To solve trigonometric equations involving $\sin 2A$ or $\cos 2A$, use the formulae $\sin 2A = 2\sin A\cos A$, $\cos 2A = 2\cos^2 A - 1$ or $\cos 2A = 1 - 2\sin^2 A$.

5. If you are asked to express a trigonometric expression in the form $k\cos(x \pm \alpha)$ or $k\sin(x \pm \alpha)$, it is essential that you use the formulae $\cos(A \pm B) = \cos A\cos B \mp \sin A\sin B$ or $\sin(A \pm B) = \sin A\cos B \pm \cos A\sin B$.

6. The tables of standard derivatives and standard integrals should **always** be referred to when differentiating or integrating $\sin ax$ or $\cos ax$.

Suppose for example that you are asked to find $\int \sin 2x\,dx$. The table of standard integrals leads you directly to the answer, $-\frac{1}{2}\cos 2x + C$.

7. Finally, remember that some formulae may appear in the reverse order.

For example, in a question involving $\cos A \cos B - \sin A \sin B$, it might help you if you looked at the List of Formulae and realised that this equals $\cos(A + B)$.

Formulae not on the List

There are many formulae, too many to list them all, which do not appear on the List of Formulae. Such formulae have to be memorised. Here are some of the more important ones.

I The formulae which appeared on the List at Standard Grade or Intermediate 2 could be useful. These are:

1. The roots of $ax^2 + bx + c = 0$ are $x = \dfrac{-b \pm \sqrt{(b^2 - 4ac)}}{2a}$.

   (This is known as the Quadratic Formula.)

2. Sine rule: $\dfrac{a}{\sin A} = \dfrac{b}{\sin B} = \dfrac{c}{\sin C}$.

3. Cosine rule: $a^2 = b^2 + c^2 - 2bc\cos A$ or $\cos A = \dfrac{b^2 + c^2 - a^2}{2bc}$.

4. Area of a triangle: $\text{Area} = \dfrac{1}{2}ab\sin C$.

5. You should also remember two important trigonometric identities:

$$\sin^2 A + \cos^2 A = 1 \quad \text{and} \quad \tan A = \frac{\sin A}{\cos A}.$$

II The next group of formulae are taken from the following Higher topics:

The Straight Line

1. The distance, d, between (x_1, y_1) and (x_2, y_2) is $d = \sqrt{(x_2 - x_1)^2 + (y_2 - y_1)^2}$.

2. The midpoint of the line joining (x_1, y_1) and (x_2, y_2) has coordinates $\left(\dfrac{x_1 + x_2}{2}, \dfrac{y_1 + y_2}{2} \right)$.

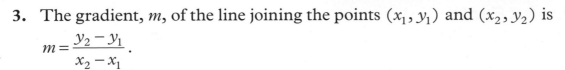

3. The gradient, m, of the line joining the points (x_1, y_1) and (x_2, y_2) is
$$m = \frac{y_2 - y_1}{x_2 - x_1}.$$

4. Lines with non-zero gradients m_1, m_2 are perpendicular $\Leftrightarrow m_1 \times m_2 = -1$.

5. The line through the point $(0, c)$ with gradient m has equation $y = mx + c$.

6. The line through the point (a, b) with gradient m, has equation $y - b = m(x - a)$.

   NOTE: Some textbooks give this formula as $y - y_1 = m(x - x_1)$.

Recurrence Relations

The formula for finding the limit of the recurrence relation $u_{n+1} = au_n + b$ is $L = \dfrac{b}{1-a}$.
(Limit exists if $-1 < a < 1$.)

Integration

The formula for the area enclosed by the curves $y = f(x)$ and $y = g(x)$ and the lines

$x = a$ and $x = b$ if $f(x) \geq g(x)$ and $a \leq x \leq b$ is $A = \displaystyle\int_a^b [f(x) - g(x)]dx$.

Vectors

1. For two vectors $\mathbf{a}, \mathbf{b} \neq 0$, $\mathbf{a.b} = 0 \Leftrightarrow \mathbf{a}$ is perpendicular to $\mathbf{b}$.

2. The distance, d, between points (x_1, y_1, z_1) and (x_2, y_2, z_2) is
$$d = \sqrt{(x_2 - x_1)^2 + (y_2 - y_1)^2 + (z_2 - z_1)^2}.$$

3. The Section Formula:

   If a point P divides a line AB in the ratio $m : n$, then $\mathbf{p} = \dfrac{m\mathbf{b} + n\mathbf{a}}{m + n}$.

In practice, it is probably easier to use a sketch with crossing lines. (See Practice Paper B, Paper Two, Question 1(a) for an example of how this works.)

Further Calculus

$$\int (ax + b)^n = \frac{(ax + b)^{n+1}}{a(n + 1)} + C, \quad n \neq -1$$

Some Useful Facts

In addition to familiarity with using formulae, it is essential that you learn the following facts.

The Straight Line

1. The gradient of a line is the tangent of the angle between the line and the positive direction of the x-axis (sometimes written as $m = \tan\theta$).

2. Make sure you do not confuse median, altitude and perpendicular bisector.

   The **median** is a line in a triangle from a vertex to the midpoint of the opposite side.

   The **altitude** is a line in a triangle from a vertex that is perpendicular to the opposite side.

   The **perpendicular bisector** of a line is a line passing through the midpoint of and at right angles to the given line.

Related Functions

Given the graph of $f(x)$, the following describes how to sketch the graph of related functions:

1. $y = -f(x)$ reflect $y = f(x)$ in the x-axis

2. $y = f(-x)$ reflect $y = f(x)$ in the y-axis

3. $y = -f(-x)$ give $y = f(x)$ a half turn rotation about the origin O

4. $y = f(x) + a$ slide $y = f(x)$ a units up (parallel to the y-axis)

5. $y = f(x - a)$ slide $y = f(x)$ a units to the right (parallel to the x-axis)

6. $y = af(x)$ stretch $y = f(x)$ parallel to the y-axis by a scale factor of a

7. $y = f\left(\dfrac{x}{a}\right)$ stretch $y = f(x)$ parallel to the x-axis by a scale factor of a

8. $y = f^{-1}(x)$ reflect in $y = x$.

Increasing and Decreasing Functions

1. A function is increasing when $f'(x) > 0$.

2. A function is decreasing when $f'(x) < 0$.

3. A function is stationary when $f'(x) = 0$.

Trigonometry

1. $\sin(-x)^\circ = -\sin x^\circ; \quad \cos(-x)^\circ = \cos x^\circ;$

$\sin(180-x)^\circ = \sin x^\circ; \quad \cos(180-x)^\circ = -\cos x^\circ;$

$\sin(360-x)^\circ = -\sin x^\circ; \quad \cos(360-x)^\circ = \cos x^\circ;$

$\sin(90-x)^\circ = \cos x^\circ; \quad \cos(90-x)^\circ = \sin x^\circ.$

2. The exact values of sin, cos and tan of 30°, 45° and 60°:

(The following results are very likely to be tested in the non-calculator paper.)

	30°	45°	60°
sin	$\dfrac{1}{2}$	$\dfrac{1}{\sqrt{2}}$	$\dfrac{\sqrt{3}}{2}$
cos	$\dfrac{\sqrt{3}}{2}$	$\dfrac{1}{\sqrt{2}}$	$\dfrac{1}{2}$
tan	$\dfrac{1}{\sqrt{3}}$	1	$\sqrt{3}$

3. In the non-calculator paper, you are likely to need to know the values of trigonometric expressions such as sine and cosine of 0°, 90°, 180°, 270° and 360°.

It may help you to access these values quickly if you refer to the basic sine and cosine graphs.

(a) The Sine Graph

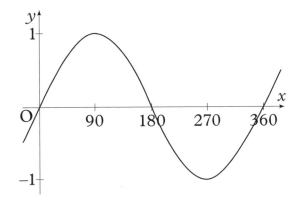

(b) The Cosine Graph

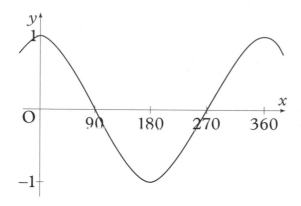

Using these graphs, you can read off $\sin 0° = 0$, $\sin 90° = 1$, $\sin 180° = 0$, $\sin 270° = -1$, $\sin 360° = 0$, $\cos 0° = 1$, $\cos 90° = 0$, $\cos 180° = -1$, $\cos 270° = 0$ and $\cos 360° = 1$.

Vectors

1. A unit vector is a vector of length (magnitude) 1 unit.

2. Remember that **i**, **j** and **k** are unit vectors parallel to the x, y and z-axes respectively

   and that the column vector $\begin{pmatrix} a \\ b \\ c \end{pmatrix}$ is equivalent to $a\mathbf{i} + b\mathbf{j} + c\mathbf{k}$.

The Laws of Logarithms

1. $\log_a uv = \log_a u + \log_a v$

2. $\log_a \left(\dfrac{u}{v} \right) = \log_a u - \log_a v$

3. $\log_a u^v = v \log_a u$

4. It is essential to remember the relationship between exponential and logarithmic statements: $N = b^x \Leftrightarrow \log_b N = x$. This can help you get started in some questions in logarithms. To help you remember it, you could note a particular example, e.g. $1000 = 10^3 \Leftrightarrow \log_{10} 1000 = 3$.

5. Remember, too, that $\log_a 1 = 0$ and $\log_a a = 1$ for **any** base.

Appendix 2

Objective Test Questions

In this section, we look at techniques for tackling objective test questions. There are **five types** of objective questions which can appear in the Higher Maths examination. These are as follows.

Question Types

1   Direct question

(Example)

A circle has equation $x^2 + y^2 + 10x - 4y - 35 = 0$.

What is the radius of this circle?

A $\sqrt{151}$

B $\sqrt{56}$

C   9

D   8

(The correct answer is D.)

2   Direct question: options in a table

(Example)

Two points A and B have coordinates $(4, d, 8)$ and $(-6, 10, e)$ respectively. The midpoint of AB is the point M $(-1, 4, -2)$.

What are the values of d and e?

	d	e
A	−2	−12
B	−2	12
C	−6	−12
D	−6	12

This is often used when there are two parts to an answer. (The correct answer is A.)

3 Direct instruction

(Example)

Find $\int (2x-1)^3 dx$.

A $\dfrac{1}{4}(2x-1)^4 + C$

B $\dfrac{1}{2}(2x-1)^4 + C$

C $\dfrac{1}{8}(2x-1)^4 + C$

D $(x^2 - x)^4 + C$

(The correct answer is C.)

4 Choice reference

(Example)

Which of the following graphs could represent the function $f(x) = \log_3 x$?

A
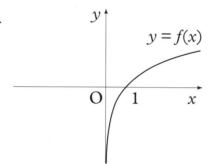

B
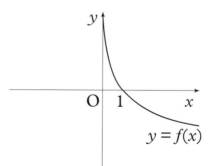

C
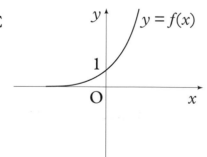

D
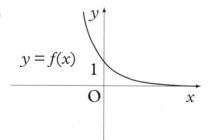

(The correct answer is A.)

5 Simple multiple completion

(Example)

Here are two statements about the equation $x^2 + 2px + p^2 = 0$, where $p > 0$, p is real:

(1) the roots are real.   (2) the roots are equal.

Which one of the following is true?

A   Neither statement is correct.

B   Only statement (1) is correct.

C   Only statement (2) is correct.

D   Both statements are correct.

(The correct answer is D).

Strategies

The recommended method to answer each of these five types of question is demonstrated and explained throughout the Answer Support for Paper One.

However, while the recommended methods to solve objective questions are explained, you should be aware that other strategies may be used in particular questions.

The following five examples show alternative techniques for solving these multiple choice questions.

1. When $x^2 + 6x + 1$ is written in the form $(x + a)^2 + b$, what is the value of b?

    A   −10

    B   −8

    C   −5

    D   10

This question can be solved quickly and fairly simply if you realise that it is based on completing the square. But what if you do not realise this?

There is still hope as long as you realise that $(x + a)^2 + b = x^2 + 2ax + a^2 + b$ (by squaring the bracket).

Hence $x^2 + 6x + 1 = (x + a)^2 + b = x^2 + 2ax + a^2 + b$

$\Rightarrow 2a = 6$ and $a^2 + b = 1 \Rightarrow a = 3$. So $9 + b = 1 \Rightarrow b = -8$.

So the correct answer is B.

This method is not the most common one. However, if you were stuck it would provide the correct answer.

2. (Here is an example in which you could use the four possible answers to solve the problem.)

Given that $(x-1)$ is a factor of $(x^3 - 5x^2 + kx - 5)$, find the value of k.

A 11

B 9

C –1

D –11

The recommended method for this question, in which k appears as an unknown in the synthetic division, is shown in the Answer Support for Paper One (Practice Paper D). However, the presence of the variable k in the polynomial could disturb some students. If so, you could replace k with each of the four possible answers in turn until you arrive at the correct solution.

$$
\begin{array}{r|rrrr}
 & & & k\!\downarrow & \\
1 & 1 & -5 & 11 & -5 \\
 & & 1 & -4 & 7 \\
\hline
 & 1 & -4 & 7 & 2 \\
\end{array}
\qquad
\begin{array}{r|rrrr}
 & & & k\!\downarrow & \\
1 & 1 & -5 & 9 & -5 \\
 & & 1 & -4 & 5 \\
\hline
 & 1 & -4 & 5 & 0 \\
\end{array}
$$

Fairly quickly you find that replacing k with 9 leads to a remainder of 0, so the correct answer is B.

This is an alternative approach to solving the problem, by trying each of the given answers to see which one works.

3. (Here is an example in which looking at the solutions first would be positively encouraged.)

The diagram shows the graph of $y = f(x)$ where f is a logarithmic function.

What is $f(x)$?

A $f(x) = \log_4(x-1)$

B $f(x) = \log_2(x+1)$

C $f(x) = \log_2(x-1)$

D $f(x) = \log_4(x+1)$

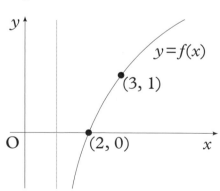

It would be difficult to find $f(x)$ without referring to the possible answers.

The best method here is to substitute (2, 0) and (3, 1) into each equation. You will find that in answer C, $f(x) = \log_2(x-1)$, substituting (2, 0) leads to $0 = \log_2(2-1) = \log_2 1$. This is true. Substituting (3, 1) leads to $1 = \log_2(3-1) = \log_2 2$. This is also true. So the correct answer is C.

(I am assuming of course that you remember that $\log_a 1 = 0$ and $\log_a a = 1$.)

4. (In some questions you might be able to rule out some of the possible answers immediately.)

    The diagram shows a circle with centre (2, 4) and a tangent drawn at the point (5, 8).

    What is the equation of this tangent?

    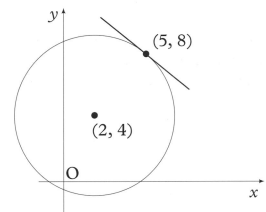

    A $y - 8 = -\dfrac{4}{3}(x-5)$

    B $y - 8 = -\dfrac{3}{4}(x-5)$

    C $y - 8 = \dfrac{3}{4}(x-5)$

    D $y - 8 = \dfrac{4}{3}(x-5)$

When you inspect the four possible answers, you will see that they are all in the form $y - b = m(x - a)$. Looking at the diagram, the tangent at the point (5, 8) has a negative gradient. Therefore before you start working, you should realise that only answers A or B are possible. Already you have cut down the possible answers from four to two. The correct answer is in fact B.

5. (This question also allows you to eliminate some possible answers at the outset.)

    By referring to the following diagram find the value of $\cos\angle DGF$.

    A $-\dfrac{16}{65}$

    B $-\dfrac{4}{13}$

    C $\dfrac{16}{65}$

    D $\dfrac{56}{65}$

    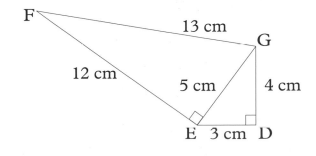

This is undoubtedly a very difficult mutiple choice question.

However, if you study the diagram, you will see that $\angle DGF$ is an obtuse angle. This means that it is between 90° and 180° and in the 2nd quadrant. As a result its cosine must be negative. Once you realise this, you can eliminate answers C and D immediately. This helps narrow your task.

There is still a considerable amount of work to do to arrive at the correct answer, but your choices are reduced from four to two. (The recommended method is shown in the Answer Support for Paper One (Practice Paper B).)

It is very important that you do not leave any blanks in the answer grid even if you have to resort to a guess, although this is **not** a recommended strategy!

Finally

When you attempt the objective test questions in your exam, be aware of the different ways of tackling each question. If the orthodox method is not working, try using the answers. You can possibly substitute them into a function or formula and see which one works correctly. Always study diagrams for clues. Eliminate impossible answers before you start doing any working. In short, use your imagination and don't be afraid to think of different approaches to questions.

Appendix 3

Hints and Tips

We have looked at objective test questions. We'll now have a more detailed look at longer questions in the examination.

Sometimes it is difficult to recognise what kind of maths should be used to get started on a question. In this Appendix, there is good advice, both general and specific, to help you get started and attain as many marks as possible.

We'll look first at **marking**. If you have an idea of how marks are allocated in common types of questions, it will focus your attention on exactly what you have to do in certain situations.

Marking

Let us consider first a common exam question in Higher Maths in which you have to find the coordinates of the turning points of a curve and their nature. We shall use a question from earlier in the book as an example.

(Practice Paper A, Paper One, Question 23)

'Find the coordinates of the turning points of the curve with equation $y = x^3 - 3x^2 - 24x + 5$ and determine their nature.' **(8)**

Solution:

$$y = x^3 - 3x^2 - 24x + 5 \Rightarrow \frac{dy}{dx} = 3x^2 - 6x - 24 = 0 \text{ at stationary points.}$$

$3x^2 - 6x - 24 = 0 \Rightarrow 3(x^2 - 2x - 8) = 0 \Rightarrow 3(x + 2)(x - 4) = 0.$

Hence stationary points occur when $x = -2$ or 4.

When $x = -2$, $y = (-2)^3 - 3 \times (-2)^2 - 24 \times (-2) + 5 = -8 - 12 + 48 + 5 = 33.$

When $x = 4$, $y = (4)^3 - 3 \times (4)^2 - 24 \times (4) + 5 = 64 - 48 - 96 + 5 = -75.$

Nature of roots:

x	$\rightarrow$	-2	$\rightarrow$	4	$\rightarrow$
$\dfrac{dy}{dx}$	$+$	0	$-$	0	$+$
	$\diagup$		$\diagdown$		$\diagup$
		max		min	

Hence $(-2, 33)$ is a maximum turning point and

$(4, -75)$ is a minimum turning point.

You can see that this question is worth 8 marks.

This is where the 8 marks are allocated:

- know to differentiate: $\left(\dfrac{dy}{dx} \right)$

- differentiate: $\dfrac{dy}{dx} = 3x^2 - 6x - 24$

- set derivative to zero: $3x^2 - 6x - 24 = 0$

- factorise: $3(x + 2)(x - 4) = 0$

- solve for x: $x = -2$ or 4

- evaluate the y-coordinates: when $x = -2$, $y = 33$; when $x = 4$, $y = -75$

- know to justify, and justify, results: (completed table showing nature of roots)

- interpret results: $(-2, 33)$ is a maximum turning point
$(4, -75)$ is a minimum turning point.

If you have a clearer idea of what is expected, hopefully it will improve your chances of attaining marks. Note the **communication** mark for realising that stationary points occur when $\dfrac{dy}{dx} = 0$. Note too the layout of the table which is used to find and justify the nature of the stationary points. Ideally your nature table should appear as shown on the previous page. It must show the labels x and $\dfrac{dy}{dx}$, and should show the slopes indicating positive, zero and negative gradients with turning points labelled max, min or point of inflection.

To find out more about how marks are allocated for particular questions in Higher Maths, ask your teacher or, alternatively, you can look up the SQA website.

Look in SQA NQ Past Papers Maths for details of how previous exams have been marked. Other useful information about the examination appears on the SQA website.

Here are some other useful hints which could gain marks for you.

Simplification

Remember always to express fractions in their simplest form. For example, if you use the gradient formula to find that $m = \dfrac{6}{8}$, it is essential that you continue and write $m = \dfrac{3}{4}$.

Similarly always simplify numerical expressions. For example, you should not leave the equation of a circle in the form $(x-3)^2 + (y-2)^2 = 5^2$. It should be simplified to $(x-3)^2 + (y-2)^2 = 25$.

Communication

Communication is most important.

1. If, say, in a vectors question, you are asked to prove that three points A, B and C are collinear, you should find the components of $\overrightarrow{AB}$ and $\overrightarrow{BC}$, show that $\overrightarrow{AB} = k\overrightarrow{BC}$, and then communicate your answer as follows:

   'A, B and C are collinear as $\overrightarrow{AB}$ and $\overrightarrow{BC}$ have a common direction and a common point.'

   Similarly in straight line questions, use 'common direction and common point' when proving points are collinear. You can use gradients in this case.

2. If you are solving a recurrence relation problem involving finding a limit, always justify that a limit exists. For example, if the recurrence relation is $u_{n+1} = 0 \cdot 4u_n + 8$ you should state 'a limit exists since $-1 < 0 \cdot 4 < 1$'.

   (As before, always simplify numerical expressions such as $\dfrac{8}{0 \cdot 2}$ to $\dfrac{80}{2} = 40$.)

3. In problems involving equal roots, as well as proving that a straight line is a tangent to a curve, always state that the condition for equal roots is $b^2 - 4ac = 0$.

4. When you understand the context of a question then give your answer in context. For example, in the optimisation problem in Practice Paper D, Paper Two, Question 8, the question asks that you 'Find the dimensions of the jewellery box with maximum volume'.

   In this case your answer should state the dimensions in words. Thus:

   'The dimensions of the jewellery box with maximum volume are 6 cm $\times$ 6 cm $\times$ 6 cm.'

General Advice

(a) It hardly needs saying that you should show all your working clearly. The earlier example in which we looked at the marking scheme for a question on stationary points shows how marks can quickly add up through clear working.

(b) Never write down multiple answers to the same question. Suppose you had two different attempts at the same question leaving the marker unsure which one was your preferred option. In this case the marker would mark both attempts and award you the lower of the two marks. So be decisive. Decide which scenario is most promising and stick to it.

(c) In some circumstances, draw a diagram to help. Here is a case in point from Practice Paper A, Paper One, Question 3:

'The straight line joining the points $(6, 0)$ and $(0, 12)$ passes through the point $(-2, a)$. Find the value of a.'

If you construct an accurate sketch by plotting $(6, 0)$ and $(0, 12)$ and extend the line joining these two points, you should be able to find that $(-2, 16)$ lies on the line and therefore $a = 16$. Thus:

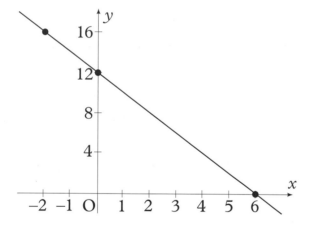

(d) In textbooks, vectors are shown in bold, e.g. $5\mathbf{a} + 3\mathbf{b}$. In your exam, you should underline vectors, i.e. $5\underline{a} + 3\underline{b}$.

(e) Many questions have more than one part. Remember that if a part begins with the word 'Hence' then it is a hint that you should use the result or formula from the previous part.

Specific Advice

(a) If you are asked to find where two graphs meet, solve the pair of equations. Use elimination to solve simultaneous equations for two straight lines. Use substitutions in all other cases, such as a straight line and a circle. In a straight line and circle question you should draw a simple sketch to help you if a diagram isn't provided.

(b) If you are asked to find a suitable domain, think of the two **impossible** calculations: dividing by zero and finding the square root of a negative number.

The latter is illustrated in Practice Paper D, Paper One, Question 13.

Here is an example of the former:

A function is given by $f(x) = \dfrac{5}{x-7}$. What is a suitable domain of f?

As division by zero is impossible, a suitable domain would be $x \neq 7$.

(c) In trigonometry, remember to check whether angles should be in radians or degrees. Angles in differentiation and integration using trigonometry will be in radians.

Remember that π radians = $180°$.

(d) A question which uses the phrase 'rate of change' will involve **differentiation**.

See Practice Paper C, Paper One, Question 15 for an example.

(e) In solving a quadratic inequality, sketch the graph of the quadratic function.

See Practice Paper A, Paper One, Question 11 for an example.

(f) A question which uses the phrase 'nature of the roots' will always involve the discriminant $b^2 - 4ac$.

(g) Questions in which you are asked to find an area, accompanied by a graph involving a shaded area, are likely to be solved by integration (you may have to work out limits first).

(h) A question in which you are given $f'(x)$ and asked to find $f(x)$ will involve integration.

(i) In indefinite integrals, always remember to write $+ C$.

(j) If you are sketching a graph, annotate the graph by showing important points such as turning points and points where the graph cuts the x and y-axes.

(k) In questions involving the auxiliary angle, you should check your answer on the calculator by substituting the same angle into both expressions. In Practice Paper A, Paper Two, Question 7(a), we proved that
$8\cos x° - 15\sin x° = 17\cos(x + 61 \cdot 9)°$.

To check whether this is correct or not, we can choose a value for x and try it in both sides of the equation.

You might choose, say, $x = 82$ and work out $8\cos 82° - 15\sin 82°$ and also $17\cos(82 + 61 \cdot 9)°$. (Allow for a small difference due to the fact that $61 \cdot 9°$ has been rounded to one decimal place.)

(l)   Optimisation questions are fairly easy to recognise. You may have to prove a given formula and then you will be asked about the maximum/minimum, largest/smallest or greatest/least value of the variable.

Remember that this type of question is just like finding stationary points.

You must differentiate, equate the derivative to zero, solve the equation, construct a nature table, and then communicate your answer in words.

This is usually a two part problem with the proof of a formula in part (a). This formula will be required to do part (b). If you feel that the proof is going to slow you down or upset you, do part (b) first using the formula and return to part (a) later. A typical optimisation question appears in Practice Paper D, Paper Two, Question 8. Study the proof in the solution and read the hint. Other proofs may be something similar.

(m)   Students often have difficulty with short 1 mark problems, and use an unnecessarily complicated method.

Consider the following problem:

'AB is the diameter of a circle with centre C. If A is the point (8, 3) and C is the point (5, −4), what are the coordinates of B?'

The 'step' from A to C must be the same as the 'step' from C to B.

By inspecting the coordinates you can see that the 'step' from A to C is 3 back and 7 down. Now use the same 'step' from C to B.

Using the x-coordinates leads to $5 - 3 = 2$.

Using the y-coordinates leads to $-4 - 7 = -11$.

Hence B is (2, −11).

No working is necessary and the calculations can often be done mentally.

A similar 'stepping process' may be used in questions involving quadrilaterals whose diagonals bisect each other, e.g. the parallelogram and the rhombus.

We finish this section of specific advice by looking at two topics which merit special mention.

Similar Triangles

Occasionally a question appears in the examination which can be solved using similar triangles. It would probably be part (a) of a two-part question and appear as a proof.

Similar triangle problems are not particularly well done by candidates at Standard Grade Maths and are not part of the Intermediate 2 syllabus. As a result you may find questions on similar triangles difficult in the Higher Maths examination.

Two triangles are similar if **either** of the following conditions are true:

1. They are equiangular.

2. Their corresponding sides have lengths in the same ratio.

(If one of these conditions is true, then the other will also be true.)

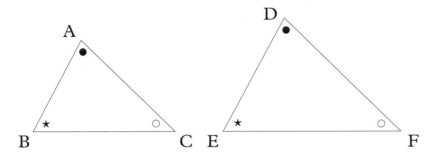

In these diagrams, triangles ABC and DEF are similar because
$\angle A = \angle D$, $\angle B = \angle E$, $\angle C = \angle F$.

Hence their corresponding sides have lengths in the same ratio:

$$\frac{AB}{DE} = \frac{BC}{EF} = \frac{AC}{DF}.$$

We can use these properties to solve Question 7(a) in Practice Paper C, Paper Two:

'The shaded rectangle in the following diagram is drawn with one vertex at the origin and the opposite vertex lying at E on the line CD where C is the point (0, 6) and D is the point (18, 0).

The other two vertices of the rectangle are the points P (0, p) and Q (q, 0).

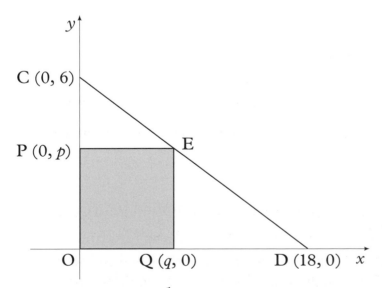

(a) Show that $p = \frac{1}{3}(18 - q)$ and hence deduce that the area, $A(q)$, of the

rectangle is given by $A(q) = 6q - \frac{1}{3}q^2$.'

(3)

There are actually three similar triangles here, triangle OCD, triangle PCE and triangle QED.

We concentrate on triangles OCD and QED.

Triangle OCD is similar to triangle QED because they are equiangular.

$$\angle COD = \angle EQD, \quad \angle OCD = \angle QED, \quad \angle CDO = \angle EDQ.$$

(The reasons are that the angles at O and Q are right angles, the angles at C and E are **corresponding** angles (F-shape) and angle D is common to both triangles.)

Hence $\dfrac{OD}{QD} = \dfrac{OC}{QE} = \dfrac{CD}{ED}$.

We can use the first of these equalities to proceed.

By substitution, $\dfrac{18}{18-q} = \dfrac{6}{p}$.

Therefore $18p = 6(18-q) \Rightarrow p = \dfrac{6}{18}(18-q) = \dfrac{1}{3}(18-q)$.

(The remainder of the proof follows using length × breadth.)

You would not need to include all the theory in your answer, but simply start by naming the similar triangles. Then go straight to $\dfrac{18}{18-q} = \dfrac{6}{p}$.

The Graph of the Derived Function

This type of question appears occasionally and causes students difficulty.

You are given the graph of a function $y = f(x)$ and asked to draw the graph of the derived function $y = f'(x)$.

Here are some key points to remember when doing this:

- a stationary point at (a, b) on $y = f(x)$ means $(a, 0)$ lies on $y = f'(x)$

- where $y = f(x)$ is **increasing**, the graph of $y = f'(x)$ is **positive** (above the x-axis)

- where $y = f(x)$ is **decreasing**, the graph of $y = f'(x)$ is **negative** (below the x-axis)

- the degree of $y = f'(x)$ is one less than the degree of the polynomial $y = f(x)$.

The obvious cases to remember are that the derivative of a quadratic function (whose graph is a parabola) is a linear function (a straight line graph), and the derivative of a cubic function (with a maximum and minimum turning point) is a parabola.

(This is clear if you think about the function $f(x) = x^2$. This is a quadratic function whose graph is a parabola. As $f'(x) = 2x$, the graph of the derived function is a straight line.)

Example 1: The diagram shows a sketch of the function $y = f(x)$.

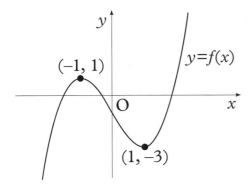

Sketch the graph of $y = f'(x)$.

Solution:

The stationary points at $(-1, 1)$ and $(1, -3)$ mean that $(-1, 0)$ and $(1, 0)$ lie on the graph of $y = f'(x)$.

As $y = f(x)$ is increasing for $x < -1$ and $x > 1$, the graph of $y = f'(x)$ is above the x-axis for these intervals. It will be below the x-axis for $-1 < x < 1$.

The graph of $y = f(x)$ appears to be a cubic function (with a maximum and minimum turning point), so the graph of $y = f'(x)$ will be a parabola.

This analysis leads to the following graph.

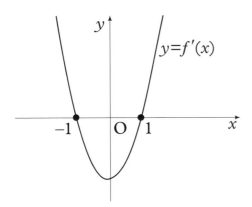

The next example deals with the graph of a polynomial function of degree 4.

(The graph has a point of inflection and a maximum turning point.)

Example 2: The diagram shows a sketch of the function $y = f(x)$.

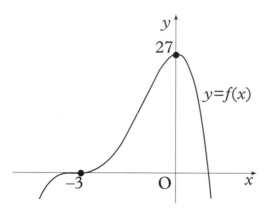

Sketch the graph of $y = f'(x)$.

Solution:

The stationary points at $(-3, 0)$ and $(0, 27)$ mean that $(-3, 0)$ and $(0, 0)$ lie on the graph of $y = f'(x)$.

As $y = f(x)$ is increasing for $x < -3$ and $-3 < x < 0$, the graph of $y = f'(x)$ is above the x-axis for these intervals. It will be below the x-axis for $x > 0$.

The graph of $y = f(x)$ is a poynomial function of degree 4, so the graph of $y = f'(x)$ will be a cubic function.

This analysis leads to the following graph.

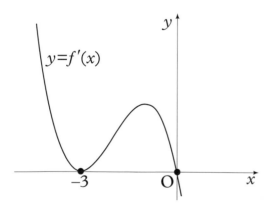

This may seem tricky, but if you learn the key points and, importantly, practise several examples your confidence will soon improve.

Appendix 4

Practice Paper Questions by Topic

In order that you may study examination style questions as you complete each topic, the following information lists each question from the four specimen papers under the main topic headings from the Higher syllabus.

KEY

A, 1, 5 refers to Practice Paper A, Paper One, Question 5.

C, 2, 8 refers to Practice Paper C, Paper Two, Question 8.

NOTE: Where a question covers more than one topic, it appears under the latter topic in the syllabus.

The Straight Line

A, 1, 3
A, 1, 7
A, 1, 18
A, 1, 22
B, 1, 3
B, 1, 23
C, 1, 1
C, 1, 21
D, 1, 5
D, 1, 21

Functions and Graphs

A, 1, 21
A, 2, 2
B, 1, 6
B, 1, 11
B, 1, 17
B, 2, 6(a)
C, 1, 12
D, 1, 8
D, 1, 13

Differentiation

A, 1, 2
A, 1, 6
A, 1, 16
A, 1, 23

A, 2, 4(a)
B, 1, 21(a)
B, 1, 22
B, 2, 5
C, 1, 5
C, 1, 14
C, 1, 15
C, 2, 4
C, 2, 7
D, 1, 16
D, 1, 22
D, 2, 8

Recurrence Relations

A, 1, 4
A, 1, 12
B, 1, 1
B, 1, 9
C, 1, 2
C, 1, 6
D, 1, 1
D, 2, 3

Identities and Radians

A, 1, 14
B, 1, 12
C, 1, 7
D, 1, 6

Quadratic Theory

A, 1, 9
A, 1, 11
B, 1, 10
B, 1, 13
B, 1, 20
C, 1, 4
C, 1, 9
C, 1, 13
C, 1, 16
C, 2, 1
D, 1, 14
D, 1, 15
D, 2, 7

The Remainder Theorem

A, 2, 5
B, 1, 21(b) and (c)
C, 1, 17
C, 1, 22
D, 1, 4
D, 2, 5(a) and (b)

Integration

A, 1, 20
A, 2, 4(b)
B, 2, 8
C, 1, 19
C, 2, 6
D, 2, 5(c)

Compound and Multiple Angles

A, 1, 5
A, 1, 17
A, 1, 24(a) and (b)
A, 2, 3
B, 1, 19
B, 2, 3
B, 2, 6(b)
C, 1, 24
C, 2, 3
D, 1, 18

The Circle

A, 1, 1
A, 1, 8
A, 2, 9
B, 1, 2
B, 1, 7
B, 2, 2
C, 1, 8
C, 1, 23
D, 1, 2
D, 1, 7
D, 1, 23

Vectors

A, 1, 15
A, 2, 1
A, 2, 6

B, 1, 4
B, 1, 8
B, 1, 14
B, 1, 16
B, 2, 1
C, 1, 3
C, 1, 11
C, 2, 2
D, 1, 3
D, 1, 10
D, 1, 11
D, 1, 17
D, 2, 1

Further Calculus

A, 1, 19
A, 1, 24(c)
B, 1, 5
B, 1, 15
C, 1, 10
D, 1, 9
D, 1, 12
D, 2, 2

Logarithms

A, 1, 10
A, 2, 8
B, 1, 18
B, 2, 7
C, 1, 20
C, 2, 8
D, 1, 19
D, 1, 20
D, 2, 6

The Auxiliary Angle

A, 1, 13
A, 2, 7
B, 2, 4
C, 1, 18
C, 2, 5
D, 2, 4